Small-Scale
HAYMAKING

Small-Scale
HAYMAKING

SPENCER YOST

Voyageur
Press

Dedication

Dedicated to the memory of Sammy Sherrill, the greatest haying buddy someone could ask for and a role model the world needs.

First published in 2006 by Voyageur Press, an imprint of MBI Publishing Company, Galtier Plaza, Suite 200, 380 Jackson Street, St. Paul, MN 55101-3885 USA

MBI Publishing Company titles are also available at discounts in bulk quantity for industrial or sales-promotional use. For details write to Special Sales Manager at MBI Publishing Company, Galtier Plaza, Suite 200, 380 Jackson Street, St. Paul, MN 55101-3885 USA

Library of Congress Cataloging-in-Publication Data

Yost, Spencer, 1961-
  Small-scale haymaking / by Spencer Yost.
    p. cm.
  Includes index.
  ISBN-13: 978-0-7603-2096-9
  ISBN-10: 0-7603-2096-9
  1. Hay. I. Title.
  SB198.Y67 2006
  633.2—dc22

                              2006006834

About the author:
Spencer Yost raises hay on a small-scale farm in northwestern North Carolina, where he lives with his wife. He has two grown children, and when he isn't in the field, he is a contracting software engineer and operates the antique tractor websites www.antique-tractor.com and www.atis.net.

On the cover: (main image) Most balers only create bales that are as consistent as the feed rate and style with which the baler was fed. Here is an example of improperly biased loading. Biased and inconsistent loading and light windrows lead to short and/or lopsided bales in most balers.
(Detail image) This picture shows proper elevator loading technique. Stay away from the bottom of the elevator because the chain will grab boots and their laces. Load from the side and away from the motor. Wait a few seconds before loading the next bale. Most motors sized properly for 14- to 20-foot elevators won't be able to raise more than two or three bales at a time without overheating. Of course, if your elevator is short, this may not be a concern.

On the back cover:
Every hay field will have a sharp corner or two that can't be avoided. Here is how you operate a mower around such a corner. Start getting ready for the turn by having your hand on the hydraulic control so you can raise the mower. As you get close, time the raising to exactly match the end of the row, and raise the mower without shutting down the mower. As you start the turn, shut down the mower. You shouldn't turn any operating implement sharply while it is operating, and turning sharply is necessary to make these turns efficient. Start the mower again and get your hand on the hydraulic control once more. Drop the mower when you reach the start of the row. If you time it right, the mower won't clog on mown hay because you dropped too soon and there won't be any un-mown hay where you dropped the mower too late.

Editor: Amy Glaser
Designer: Chris Fayers

Printed in China

CONTENTS

FOREWORD

There is something unique about the experience of hay-making. The process evokes a sense of nostalgia, a sense of accomplishment, and creates a relationship to the land that very few other farming endeavors engender. Even a large haymaking operation operates many smaller crews with reasonably sized groups of equipment and people. Haymaking typically stays tied more closely to the environment it operates in and the machinery used are not ground thumping monsters, but equipment we can relate to in size and complexity.

An acquaintance who raises 300 acres of orchard grass put it this way:

"There is no difference between my grandfather's 30-acre haymaking operation and mine. We cut, we dry, we bale, and we stack. Sure my equipment is a bit nicer and safer and we do it more often, but otherwise there is no difference. Except maybe he didn't have a hay elevator and I do. My brother got the other half of the farm and raises corn and soybeans. He is a slave to bankers and to equipment he can't understand or work on. He'd switch but he's trapped and he'll have to farm the high-revenue, low-margin stuff for several years until he can get out from underneath it."

I was thinking about what my acquaintance said while I was making a late fall hay cutting last year as my neighbor was combining corn. He had just bought a new combine and was rushing as he always did to get his corn to the dryer. The combine was very nice and had to have cost much more than my house did 10 years ago. He was oblivious to his surroundings and talking on a cell phone. I imagine he was talking to the crew at the dryer, suppliers, banker, or someone like that about a topic that just couldn't wait.

That is when the deer jumped. My neighbor never saw it.

These are the rewards of the small-scale haymaker. Sure, I'll talk a little about money and economics in this book because it is important that we don't squander resources, but the important set of books is the books of our lives. When reconciling that final set of books, remember that your hay-making operation contributed mightily to those balance sheets. You will be richer for knowing some deer individually by their markings. You'll be richer for the adventure of mowing a hornet's nest and you'll be richer for knowing exactly how the air feels as it settles when the sun goes down.

The combine continued moving past the deer like nothing had happened. Like the world didn't just open up and stand there in all its glory, begging all present to look up from their work and marvel. Heaven lowered its finest treasure today and I watched someone miss it. Thank God it wasn't me.

INTRODUCTION

This book will take you through the process of developing a hay field, harvesting it, maintaining it, and developing a farm plan for the future. Along the way you'll learn about the equipment you'll need, how to find and buy it, and the procedures for operating and maintaining it. I will also discuss selling hay, marketing it, storing it, and will help you decide what type of hay to grow.

I wrote this book as if you were raising hay to sell. I did this is because virtually everyone who raises hay sells hay, even if they are mainly raising hay for their own animals. Likely you will have some hay left over to sell and you may not have animals forever. While I tried not to mention brand names of equipment, I found that difficult, so I have named brands when needed as part of the example or where a particular brand, such as New Holland, is so overwhelmingly popular it is impossible to avoid.

I made a few other assumptions. I assume you are comfortable around equipment and have a typical and reasonably complete set of tools. I assume you have a farm truck of some type and trailers that will hitch up to the truck for transporting the equipment and/or hay if needed. I do not cover how to trailer equipment in this book. Check out the *Antique Tractor Bible* for that information. I also don't discuss shop repairs and procedures except in passing as too many other MBI Publishing Company titles cover these topics really well.

I also assume you have some experience with farming in general. Haymaking can be done by anyone who is devoted to the idea, so don't let a lack of experience deter you. Simply double your efforts to train and educate yourself and gain experience through helping other haymakers beforehand. Buying this book is a great start.

Just remember, no bale of hay is worth life, limb, or property. A barn full of hay is not the goal when haymaking. The goal is to go home to your family at the end of the day with all your limbs and your life. Safety is the most important goal. Please research and carefully evaluate all pieces of equipment you are using for safety risks and have experienced people help you operate any equipment that you haven't operated before. For example, if a baler is foreign to you, ask an experienced neighbor to give you a hand

Part of haymaking is having shop space to call your own, even if you aren't planning on doing any repairs yourself. Being able to store backup parts, fuel, oil, and hardware is important. Your shop doesn't have to be elaborate, and improvising, such as using something like this library card catalog cabinet for fastener storage, can be a way to save space and stay organized and stocked.

The haymaker's world is filled with flammable fuel, ignitable hay, and hot equipment that creates sparks. The danger of fire is very real and steps must be taken to prevent it. You should have fire extinguishers, such as this one, on your tractor and in your shop and barn. Two important safety items to have in the shop and the field and barn are a complete first aid kit and a cell phone for emergencies.

A finish mower (right) and a brush cutter (below) are two handy implements for keeping wood and fence lines trim and neat, but we don't mow hay with them. They chop and clump the hay, which makes it impossible for the hay to dry and bale.

Haymaking isn't just for middle-aged men. Many younger folks interested in horses and older folks looking to fill retirement days get into raising hay on a small scale to supplement income, save on hay expenses, or for the fresh air and exercise. Just remember that this can be a dangerous pastime and shouldn't be entered into lightly like you might some other new hobby.

This was how hay was baled in the old days. A hay press is operated either by a tractor or a large stationary engine. The size of the belt pulley indicates the hay press is from the days of early gas- or late steam-engine tractors. The hay was forced down through the opening and compressed in a bale chute, much like current balers. However, farm hands would feed lengths of wire through the hay while it was in the bale chamber and twist them off to make a wire-tied bale. There were no knotters on these early hay presses.

In some climates, one of the early decisions you'll have to make is whether you will irrigate your hay. Small-acreage hay farms can irrigate using portable pumps powered by the PTO shaft of the tractor. This is an example of an irrigation system that would keep 5–10 acres irrigated with the right sprinklers.

the first time. In short, never be afraid to admit ignorance; and remember that inexperience is curable—arrogance and pride are usually self-limiting. Last but not least, please get to know your agricultural extension agent very well. The extension folks there are paid to help farmers succeed. They are very knowledgeable and can help extensively in all aspects of your operation.

I also want to thank a few folks. As usual, I want to thank the men and women of my website, ATIS.net. Over the years they have been a source of inspiration and friendship that I value. I also want to thank Leinbach Equipment in Winston-Salem, North Carolina, for allowing me to take a few pictures of implements I don't use but wanted to feature in this book. Thanks also to Jake Brewer for being a good friend and a great source of haying knowledge. Many thanks also to the good folks at the Forsyth County Cooperation Extension Service in North Carolina for all their help.

Enjoy and stay safe!

The first real expense for any new operation is going to be the tractor. The smaller Massey Harris will not operate a baler or a mower-conditioner, but it would make a great raking tractor. A larger tractor, such as this Allis Chalmers 200, isn't nimble enough and easy enough to work on for most of our small haymaking operations. In general, look for a tractor that generates about 30-45 horsepower on the PTO shaft to safely operate most hay equipment. Pick up a copy of the Nebraska Tractor Tests to research the performance parameters of any model you might be considering. It would be impossible to list all of the tractors that would make great small acreage tractors, but I will mention one

choice that crosses most people's mind: the Ford 2N, 9N, and 8N tractors. While in certain circumstances and in combination with certain types and styles of implements, they may operate successfully as power units for your operation, you should get a slightly heavier, more powerful tractor. The Ford 600 series or the Massey Ferguson Model 65 are two tractors nearly identical to these tractors in configuration and feel, but they are heavy enough and strong enough to handle the work.

CHAPTER 1
HAYING FINANCES

Small-scale haymaking, while never a wildly profitable endeavor, can be quite lucrative in the right market. In most areas of the country, the "right market" means growing high-quality hay for horse owners. Also, growing for yourself can quite profitable as well, since a penny saved is a penny earned. Putting bales in the loft that cost you 50 cents to make sure beats putting bales in the loft that cost you $3 to buy, plus the aggravation and expense of pickup/delivery—making your hay has the added benefits of being able to create hay exactly as you like it and baling in sizes and tensions that suit you. You may even have enough left over to sell that will help pay for some of the expenses.

THE ECONOMICS OF BUYING EQUIPMENT

When you start out, I encourage you to price new equipment because I want you to understand how utterly expensive it is and how completely unfeasible it is to raise hay on a small scale with equipment that costs more than a few thousand dollars. Therefore, the greatest challenge in setting up a new haymaking operation is finding serviceable and reliable field-ready pieces of equipment for next to nothing. I admit that this is very difficult. I describe some strategies for shopping for equipment in a later chapter, but I did want to cover a basic strategy here.

The first approach is an outright purchase of used equipment. This option should be pursued if at all possible.

Above and previous pages: One of the decisions you'll face when buying used equipment is how old is too old? Here are two baling rigs: the Oliver 77/IH model 46 baling outfit is from the 1950s (top left), and the Ford/New Holland outfit is from the 1980s (bottom left). While the Ford/New Holland has a three-point hitch, a ROPS (roll-over protection structure) and probably better parts availability, it has very little else to recommend it over the Oliver/IH rig. Both perform beautifully in the field and both balers will have roughly the same rate of broken/deformed bales if set up properly. The question comes down to this: Is the added money spent on the newer equipment worth the lesser risk of breakdowns and ruined hay caused by those breakdowns? Depending on the time of year and market, the cost difference between the two rigs is between $8,000 and $12,000. If you decide on very old equipment, are you comfortable working on it and maintaining it yourself, because repairs will be expensive to hire out? These are the questions you must ask and honestly answer when deciding on equipment.

Owning your own equipment is nearly mandatory if you want the flexibility of harvesting hay when you want to.

Another approach to acquiring affordable equipment is to cost share the item. For instance, an acquaintance makes hay on his five acres and also makes hay on his neighbor's five acres. My acquaintance bought half the equipment and his neighbor bought half. The neighbor then pays my friend a flat fee for operating the equipment and loading the hay in the barn. This arrangement made the equipment affordable and they both have hay—a win-win situation.

Another approach is to buy and sell equipment. If you are handy with a wrench, a welder, and a paint gun you can use equipment as you fix it up and paint it, sell it for a profit after a year or two of use and refurbishment, then put all the proceeds into a nicer piece of equipment. This is a great way to trade skills for cash without a lot of upfront investment.

SOME THINGS TO CONSIDER

One primary problem with older, used haymaking equipment is the fact it does not have modern safety systems. Operator guards, such as the guard over a power take-off shaft, are often nonexistent or were once installed and are now missing. Fabricating or locating these guards can be time consuming, costly, or both. PTO (power-takeoff) shafts on some older tractors and implements may be nonstandard and need adapters and bushings to make them work. Also, many older

Older tractors work well for most small-scale hay farms. This is the back end of a John Deere B, which has a nice swinging drawbar for drawbar work, such as pulling a small hay trailer with 40-70 bales of hay. An over-running clutch that adapts from the older thinner PTO shafts seen here to the newer larger PTO shafts is all you need to prepare it for PTO work, such as operating a hay rake.

tractor operation safety rules should guide your approach to all the other safety rules you may encounter in this book, manuals, training classes, and/or friends, family, and neighbors.

- Never go near a shaft, implement, engine, or piece of equipment that is moving, spinning, or rolling.
- Never get off the tractor seat unless all attached equipment has fully stopped and the parking brakes are set.
- Never adjust the tension of any piece of equipment while it is running.
- Never assume that blind spots in front of tractors and behind implements are clear.
- Always know the location of bystanders.
- Fatigue and sense of urgency are two of the attributes most associated with agro-industrial accidents.
- Never risk anything for a $3 bale of hay.

SAFETY ISSUES

Baling hay is a very dangerous job. All the equipment is sharp and heavy, and most spin fast when being operated. This is nothing at all like dragging a set of cultivators behind the tractor or mowing your yard with a modern lawn tractor with a full complement of safety shields. While I am fortunate enough to not have a nick or scratch to show for the haymaking I have done, I don't let that lull me into complacency. I have many stories I hope will give you pause to reflect the seriousness of the safety issues around haymaking. I'll share a couple in the hopes that it will sharpen your risk-awareness skills and maybe motivate you to read my list of core safety rules.

Story number one involves a good friend of the family who lost her fiancé to a baling accident. He died while trying to adjust or clear (the paramedics couldn't tell from the remains) his baler while it was running. He forgot the cardinal rule: If the PTO shaft is turning, stay on the operator's seat until the implement has completely stopped. Ignore this rule at your own peril.

The next story involves a neighbor who was severely injured by a rotary cutter. He stepped on the cutter's housing while it was running, not realizing that the housing had rusted to the point that it couldn't hold his weight. Even though he had stepped on the housing a thousand times before, he couldn't step on it the 1,001st time. Again, if he had observed the cardinal rule of staying on the tractor until the implement had completely stopped, he would still have his foot. I guess my message is clear. No hay is worth even a small injury. Please stay alert, stay focused, try to anticipate safety concerns and issues, and never be afraid to shut everything down immediately if you ever become alarmed or concerned for any reason.

tractors do not have over-running clutches built into their PTO shafts. Make sure you add one to your tractor if it doesn't have one. This is a very important safety precaution.

In short, haymaking can make excellent financial sense in modern agricultural settings if you are aware of, and willing to make, the trade-offs they require. Just remember to retrofit or repair your tractors and implements with any safety systems needed to protect life and limb.

CORE SAFETY PRINCIPLES

Like all activities, there are a few safety rules that take precedent over all other safety rules. These seven haymaking and

Here are two photos of the same shop. The first shows one end with just enough room to service a tractor or implement. The second is an area for bench work. It includes a heater (brown appliance in the background) for the cooler days, and a small sandblaster and parts cleaner can be seen on the right. While these areas are small and spartan, they certainly work perfectly for a small-scale haymaking operation.

STORAGE, SHOP, AND TOOL REQUIREMENTS

Storage Requirements

A haymaking operation requires quite a bit of storage. There is the hay to store, of course, but the equipment must be stored, too. Consider this: 10 bales of hay in a stack five layers high (the size a normal-sized person can comfortably get bales off of and onto) requires nine square feet of floor space. That means 1,000 bales of hay require at least 900 square feet of floor space in a loft at the minimum. Usually, it will be more like 1,000 square feet; even if it is packed tightly. These numbers seem small except when you realize this is all the space available in the loft of a 24x40 foot barn!

If you are considering storing your equipment outdoors, let me outline just a few reasons why you should reconsider. Wooden hay wagons, balers with intricate knotting mechanisms, and rakes with their arrays of bearings and bushings all require cover if they are to last for more than a few seasons. Mowers with sickle bars will rust to the point of seizing if they are left out in the elements. Belts do not last if exposed to sunshine for any length of time. Of course, all the paint designed to protect metal fades and then fails to protect the sheet metal of the tractor and implements. Safety curtains also do not stand up well to long-term exposure to the UV rays of the sun and the plastic on implements, such as the PTO shield, become brittle and break easily. All equipment should be housed indoors

The tools needed for small haymaking operations are pretty straightforward and obvious, but choosing which tools actually requires an extra moment of thought. For example, when choosing hand wrenches, which do you choose? No one needs or can typically afford a full set of four styles. The second from the top, a ratcheting combination wrench, is my favorite but is expensive. The open-ended wrench at the top is handy, but it tends to round off bolt heads that are stubborn. The double offset box end is nice, but it should be your last choice because it is rarely indispensable. The bottom wrench, the combination wrench, is the best all-around wrench to buy.

if at all possible. At the very least, have adequate tarps on hand for covering these implements.

Ideally, you would have access to a traditional two-story, gambrel-roof barn. This would have a loft for the hay above and an area below for equipment storage. The size mentioned above, 24x40, is probably considered an ideal size for a small hay operation. Like two-story homes, two-story barns are cheaper to build per square foot than single level and the loft keeps the hay well away from the humidity of ground level. Or if you decide to house animals in the lower half, feeding is as easy as tossing a bale or two down from the loft.

However, I am keenly aware of how expensive it is to build a barn, and most of us, if we don't have this type of building already available, don't have means necessary to justify the expense of building one. Therefore, you will have to make some compromises and come up with some creative storage solutions. In my case, I had a few sheds that were on the property when I bought it. I built a lean-to (a small, open-sided extension built to the side of a structure) off of one of them. I keep my equipment in the sheds and lean-tos and borrow loft space from a neighbor for my hay. To pay him "rent" for this space, I cut hay for him on a small field that he owns.

Shop or Repair Shed

The discussion so far hasn't taken into account the space you need for working on equipment. Small-scale haymaking requires used equipment. There is no way on this green earth that hay farms under 300 acres can justify purchasing

Toolboxes like this are a great way to make sure your tools are well cared for and stored properly. However, they don't make the trip out to the field very well so I have three small toolboxes I fill from the large shop toolbox at the start of every haying session and keep in the truck or staging area.

While you might never need this particular tool, it exemplifies the haymaking mindset you need to have when buying and acquiring tools. You have to constantly think of tools and equipment being in remote fields with no power and limited resources around you. For example, I have a manual drill that you can use in the field without any power. As you turn the crank the drill applies pressure to the bit, in addition to spinning the bit. This particular drill is an antique and I treat it gingerly, but there are more modern examples of these types of drills that you can pick up at yard sales and auctions.

Lifting and stabilizing a tractor requires a full set of jacks, cribbing (the wooden pieces), and a hydraulic jack. You should buy the largest jack you can afford. Even if you don't need the extra weight rating of the large stand, the extra height of the stand almost always comes in handy. There is also a peace of mind and wisdom in working under a jack stand that is over-engineered.

Safety is paramount and one way to prevent one common serious accident is an over-running clutch. The PTO shafts of older tractors are directly driven from the main shaft of the transmission or differential input shaft. This means that any implement, especially brush cutters and balers, that store tremendous amounts of energy when running must be coupled to the tractor so the implement doesn't continue driving the tractor forward, even if you have pressed in the clutch. That is what the over-running clutch protects against. You must get one unless you know for a fact that your tractor has its own provision to prevent over-running.

new equipment of any description. Since we have to use used equipment, we must also count on having to make the occasional repair. While simple repairs can be done with a tool box outside in the sun, others require indoor space. For example, repairs may be time consuming and happen over a period of several days or may need to be performed at night. Some repairs require care and cleanliness to carry out, something you really need indoor space for. In addition, the lighting needs and occasional power tool requirements of these repairs require a repair area with electricity. If you can work on the equipment where you store it, great. But if you can't get electricity to where you store it, you'll need to come up with a different plan. or plan on borrowing a work space when needed.

The typical shop building is 12x20 feet with a concrete floor and one open side (if you don't have to worry about theft). Install a door (roll-up or swing-out) if you need to lock the building. A string of fluorescent lights hung above and outlets along one wall could easily be serviced by a heavy (20 amp or better) house circuit run out to the shop building. This saves you from the expense of a separate electrical service entrance. Of course you need to consult your local building code before starting this project. A concrete floor is highly recommended. Soil floors are dusty, damp, and cold to lie down on while working on equipment. Of course some fixtures and

tools, like hoists and jacks, have rollers and require concrete to be even the least bit convenient to use. Jack stands have to be set on a foundation of wood if you don't have a concrete floor. In short, if you can afford it, you should try to make sure any building you build or decide to use as a shop has a concrete floor. Water and phone are luxuries. An alarm might be necessary since your tools and equipment are valuable enough to be stolen, especially if theft is common in your area.

Tools, Repairs, and General Care

Like all pieces of equipment, haying equipment and the tractor that powers them require maintenance. Determining the best schedule for maintenance is best done by consulting the owner's manual for your equipment. These can be purchased from various sources, including the Internet. I have a complete operator's and service manual for every piece of haying equipment I own except my mower-conditioner (I only have the operator's manual), even though none of them are less than 40 years old! The appendices found in this book will help you locate these sources. A service or operator's manual will tell you many important things, such as how often to lubricate components and what style and type of fluids to use (this is particularly important for hydraulic systems).

Necessary Tools and Equipment

You don't need a huge collection of tools to maintain hay equipment, and you don't have to acquire them all up front. You are more likely to purchase good-quality tools if you buy a few every time you make hay and run across the need for them. And make no mistake, quality does matter. Homeowner grade and/or cheap tools just don't work here in the hay field and you shouldn't even try if you respect your sanity. Either they will fail to do the job, destroy parts in the process, or simply don't last for more than a few uses. Top-quality tool manufacturers such as Snap-On and Mac should be seriously considered if you can afford them. Craftsmen works well, too.

Here is a list of the basic tools you will need:
- A ½-inch drive socket set with about 15-25 sockets. The socket sizes should range from ⅜ inch to as far past 1 inch as you can find.
- One socket wrench called a breaker bar (a long-handled socket wrench that does not ratchet). This tool is very important in removing the stubborn bolts you are bound to run across. This tool can suffice as your only socket wrench, but for convenience you should also buy a ratcheting socket wrench if you can afford it.
- Combination wrenches in sizes that range from ¼-inch to 1 inch or more.

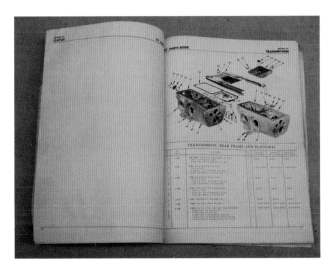

When servicing your tractor and implements, the first tool is a good set of manuals. A reprint, like the IH 46 baler manual, right, is usually just the ticket, but the reproduction quality of these aftermarket manuals can be poor and make part numbers and small instructions tough to read. This is a particularly good reprint from Binder Books. An original manual, like the parts manual seen here (left) for the Oliver 77 tractor is better, but these manuals for older tractors can be tough to find and may be considered collectible and a bit expensive. A parts manual is always a great addition to your shelf of manuals and shop reference materials. I refer to one when servicing or repairing a tractor as often as the service manuals because of the exploded diagrams found in them.

- One set of locking pliers, often known as Vice-Grips (a brand name of locking pliers).
- A 4-inch adjustable wrench and an 8-inch adjustable wrench.
- Typical common hand tools such as screwdrivers, chisels, etc.
- A grease gun.

Field Tools

In addition to the tools you will use in the shop, there are a few other tools you will need in the field. Some implements require specialized tools that should be kept on the tractor or with the implements. If you use any equipment that uses peened rivets, such as a sickle bar mower, you should have a rivet punch and a ball-peen hammer. Leverage bars and pipes are handy when moving and positioning implements. A small and inexpensive set of hand tools kept in the tractor is a real timesaver when the tractor requires attention in the field. You should also consider buying a second grease gun to be stowed on the tractor. Implements just never seem to make their way to the shop for maintenance and tend to get greased in the field.

Another set of tools you should think about acquiring are ignition/electrical tools. You will find that the majority of running problems involved with tractor engines are traced to ignition and electrical problems. Tools such as feeler gauges, a multimeter, and tachometer/dwell meter are indispensable in determining the causes of your electrical/ignition problems. Precision measuring tools are some-

times used in adjusting a baler's knotting mechanism, and the play of a bearing assembly may require calipers, a spring scale, or micrometers.

The last category of tools includes the "bottom of the box" tools. These tools always seem to work their way down to the bottom of the tool box and consist of punches, drifts, cold chisels, gasket scrapers, hacksaw blades, and the like. They are incredibly handy to have. Most of these tools can be bought as needed, although some are inexpensive enough that you should buy them whenever you have the opportunity.

GREASING A ZERK FITTING

Grease is applied to a lubrication point through a one-way greasing valve called a "zerk" (which is a trade name) or "grease" fitting. It allows the grease to enter but not leave. Steering components have these types of fittings. Knowing where all these fittings are on your tractor and implements is important. Applying grease is probably the first maintenance procedure you will perform. This is where a copy of your tractor's owner's manual will come in very handy. Most manuals have a drawing of the tractor, which outlines every grease fitting and the type of grease to use if the fitting does not take ordinary grease.

To apply grease, first clean the fitting completely and then fit the tip of the hose of the grease gun over the grease fitting and pump the grease gun. The hose, if pressed on fully and kept straight while greasing, will not come off while you pump grease but will be loose and can be easily

When loading hay on a trailer, work in a stepped pattern to minimize the amount of time you are walking on hay. Here the hay is stacked as high as it can before the next stack is started. In the beginning, many new hay farmers tend to build an entire layer on the entire floor of the trailer, then start the second layer. This requires walking on hay a lot, which tires you faster and increases the risk of falling from the trailer.

removed when you are not actually pumping grease.

I am often asked exactly how much grease should be applied to the assembly being greased. The dilemma is that too much grease is either wasteful or even potentially harmful (seals can be pushed out or damaged and components can overheat), and too little grease won't lubricate the tractor. Unfortunately, there is no set procedure for the right amount, but the following rule of thumb is pretty accurate. In most cases you are looking to pump enough grease to see a small amount escape from the joint you are lubricating or when you stop hearing air bubbles popping. In some cases, err on the side of too little. Too much grease in some high-precision assemblies creates high internal pressure during heat-up and can cause failure in the component. While this case is rare in haymaking, it does apply to universal joints found on PTO shafts. Remove excess grease from the zerk fitting from the joint immediately after you grease the fit-

ting. Grease traps dirt and moisture, and any excess helps to introduce these contaminants to the component you are trying to protect.

A QUICK NOTE ON FASTENERS

Replacing a bolt isn't as simple as finding one the right size and using it. Haying equipment is full of special bolts. For example, plow head bolts are used in bale chutes, shear pins (which are really bolts most of the time) are used on baler flywheels flanges, jam nuts are used on most tedder/rake tines, and so on. Your local farm supply store can be of assistance and you should consult experts there any time you pull a fastener that doesn't look familiar. You should get into the habit of looking on the top of every bolt you are about to replace. Identifying the hardness of a modern bolt is simple: Count the radial slashes on the head of the bolt and add two. That is the hardness. For example, a grade 8 bolt will have 6 radial slashes. The exception is: If there are no radial slashes, the hardness is either undetermined or is a grade 3 bolt. On older equipment, hardened steel bolts are dark steel and will usually scratch a modern grade 5 bolt. Otherwise assume it is a nonhardened bolt similar to a modern grade 5. Just remember to never substitute hardened bolts for shear pins. Equipment will be destroyed and life and limb may be in jeopardy.

STUCK PARTS AND "UN-STICKING" THEM

If you spend enough time around hay equipment, you will one day run across a bolt, screw, or nut that simply will not come off. Using more force will only break it off, and you don't want that to happen. Soaking the bolt with penetrating oil often works well. If giving it a few days soak doesn't work, then try applying heat. Heating the bolt will work because the heating and cooling of the bolt expands and contracts the bolt, which breaks up the rust in the threads and under the head of the bolt. A propane torch will often generate enough heat, but you will usually need an oxy-acetylene welder to generate enough heat to expand the fastener. If you don't have welding equipment, then you will probably have to drill out the bolt. But if it is a hardened bolt, drilling probably won't work. You will have to find someone with a gas welding outfit to help you heat the bolt enough to loosen it or to actually melt the nut or bolt.

TRANSPORTING AND WORKING ON BROKEN EQUIPMENT

Moving, lifting, bracing, and blocking/cribbing equipment is just part of haymaking. The time will come when an axle breaks, tires blow, and drive chains break. All of these problems lead to difficulties moving and repairing the equipment. It is very important to be especially careful here. There is no substitution for common sense and only you can tell if a piece of equipment is safe to work on or move.

Using a hay elevator carries two significant risks: snagging clothing on the conveyor chain and being unable to free a gloved hand from under the twine of a bale before you get pulled off balance. While these accidents usually just have an embarrassing ending, occasionally they end with injury and death. Here Elisha shows careful loading that keeps clothing away from the chain and particular attention is paid to freeing her hands and gloves from the bale.

It is tempting for folks new to operating tractors and equipment to think that an innocent clump of hay like this a foot or two away from any moving part would be safe to remove with the implement running, but they would be wrong. It begs the question: Why risk it? Shut down the tractor and implement every time. While this seems obvious, earlier in this chapter you read where I lost an acquaintance doing that same exact thing. Rural hospitals are filled with charts of people who did the same thing. Always shut down the implement.

For example, tedders are precariously balanced by design and may tip over when you lift them with a jack. Tines and teeth will create severe puncture wounds even if you have a very minor accident while transporting or repairing a piece of equipment. In short, be very careful and follow all normal shop precautions.

Elevating a tractor isn't like jacking up a car. The two primary concerns surrounding lifting a tractor are instability caused by the unusual front end configurations and the sheer weight of the machinery. Another variable is lifting condition, since the lifting may be done while the tractor is out in the field. Three- to eight-ton bottle or floor jacks are the best types of lifting equipment. Three-ton jacks should be used with the smaller tractors and the six- to eight-ton jacks should be used with larger tractors. I use floor jacks when I am on concrete and bottle jacks in the field. Using bottle jacks in the field require heavy wood timbers as a platform for the bottle jack. Place these foundation timbers directly on solid earth—do not place them on mud, rocks, etc. The timbers must be level also.

Securing the front end is the first step in lifting a wide-front tractor. The front axle of wide-front tractors pivot on a single point so that the tractor's front end rocks in relation to the front axle. This single point makes the tractor

unstable when you lift the tractor, especially if you lift the entire rear end at once. To secure the front end of a wide-front tractor, fasten sturdy blocks of wood between the front axle and the structural member of the tractor. Fasten the blocks tightly so they will not work loose. Narrow-front tractors are stable on level ground if only one rear wheel is lifted at one time and it is not raised very high. But check the front wheels as you lift it.

After lifting your implement or tractor, there are two rules to follow. Never get under an implement or tractor that is being held up by hydraulic jacks. The hydraulics can fail or leak down. You must support the tractor with cribbing, support, or floor jacks. These jacks lock into position and will hold their rated weight indefinitely. You should always have help nearby in case something goes wrong.

Cribbing is using strong, large wooden blocks as structural supports under the tractor to hold it off the ground. To crib, use oak blocks that are 6x6 inches or 8x8 inches and cut to handy lengths. Cribbing sections that are 2 feet in length are about right. For smaller tractors, 4x4-inch pine blocks are sufficient. Wood is used because it is strong, inexpensive, and gentle to the tractor; it conforms to slightly uneven surfaces and has a certain amount of friction between the pieces of cribbing and the tractor.

CHAPTER 2
PLANTING

Establishing a hay field is one of the most exciting aspects of making hay. There are some big decisions to make and some challenges to face. There are four primary decisions you have to make initially:

- What time of year should I plant?
- What amount of fertilizer and lime should I use?
- What type of hay?
- What type of planting method?

Of course these all come after some of the more obvious decisions, such as available acreage, the shape of the field, irrigation, etc. These decisions will lead you to new ones. For example, discussions on types of hay with the local agricultural extension agent might reveal that that you need to irrigate or spray for pests. These in turn have their own unique set of challenges and decisions. Keep your mind open and be thorough in your initial investigations so you have the opportunity to run across all possible implications

and options. This initial investigation should include discussions with experienced haymakers in your area in addition to your agricultural extension agent. I have seen many hay operations running at 50 percent of potential because the owner skipped right to discing and planting just as soon as they decided what kind of grass they wanted to plant. Like anything else in life, the results are only as good as the planning. Haymaking is no different. Great hay fields are planned and are no accident.

Contact your local agricultural extension agent to determine the best time to plant in your area. Most agents

If your tractor doesn't have a three-point hitch but your implement requires it, you can use what is called a three-point adapter or T-bar. While this doesn't work for implements that have a PTO shaft, it works great for implements, such as this plugger, a three-point disc harrow, and more that don't have a PTO shaft.

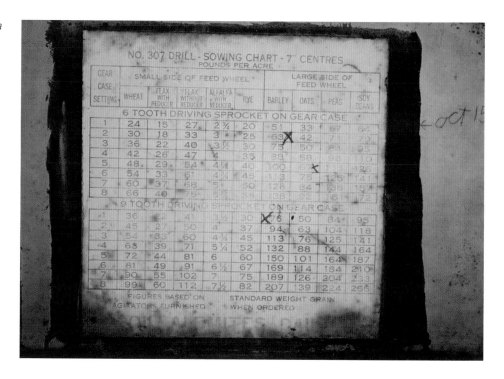

When planting your grass seed using a grain drill, follow your drill's recommendations for wheat. You can restrict the flow of seeds further by placing duct tape over a portion of each seed gate.

GEAR CASE SETTING	SMALL SIDE OF FEED WHEEL					LARGE SIDE OF FEED WHEEL			
	WHEAT	FLAX WITH REDUCER	FLAX WITHOUT REDUCER	ALFALFA WITH REDUCER	RYE	BARLEY	OATS	PEAS	SOY BEANS
6 TOOTH DRIVING SPROCKET ON GEAR CASE									
1	24	15	27	2¾	20	51	33	67	62
2	30	18	33	3	25	63	42	71	72
3	36	22	40	3½	30	75	50	84	95
4	42	26	47	4	35	88	58	98	110
5	48	29	54	4¼	40	100			126
6	54	33	61	4¾	45		75	125	141
7	60	37	68	5	50	126	84	138	157
8	66	40	75	5½	55	138	92	161	172
9 TOOTH DRIVING SPROCKET ON GEAR CASE									
1	36		41	3½	30	75	50	84	95
2	45	27	50	4	37	94	63	104	118
3	54	33	60	4½	45	113	76	125	141
4	63	39	71	5¼	52	132	88	144	164
5	72	44	81	6	60	150	101	164	187
6	81	49	91	6½	67	169	114	184	210
7	90	55	102	7	75	189	126	204	233
8	99	60	112	7½	82	207	139	224	256

FIGURES BASED ON AGITATORS FURNISHED. STANDARD WEIGHT GRAIN WHEN ORDERED.

Keep an eye on the grass seed as you drill, because even the slightest bit of trash in the seed will prevent the seed from flowing. At the end of every row I have a helper who agitates the seed with his hands to help keep the seed level and flowing.

will advise late summer, but some areas that experience an early hard frost in the fall or frequent fall droughts need to plant in the spring. The vast majority of the areas should plant in the late summer. Planting in the summer provides many benefits. A warm ground provides excellent germination rates, and the plant will be dormant for the winter before any aggressive growth takes place, allowing the plant to form a full and complete root system over the winter.

Ideally you should also wait until a few later summer rains have improved the moisture content of the soil to improve workability of the soil.

Fertilization and liming rates can often be accurately predicted by other farmers, fertilizer/lime spreading companies, and your agricultural agent without any soil tests. However, I fully encourage you to have soil tests taken to determine the exact rate of these two soil additives.

An old, inexpensive grain drill is all you need to plant a hay field. This Massey Harris from the 1940s or 1950s is what I use. The steel wheels float on soft soil well, the drill has enough spring tension for planting grass in tilled soil, and the drills are spaced close enough. In addition, I can plant about 10 feet at a time, which makes fairly short work of the planting process.

SAMPLING YOUR SOIL

Sampling your soil and reading these tests are easy. First, soil samples have to be collected from your field. The idea is to gather one sample from each part of your field that is similar. For example, the high parts of my neighbor's field are slightly sandier, but the low parts have soil that is heavier and contains more silt. Therefore, he creates two samples of dirt and has them analyzed separately. He then implements the recommendations from those reports for each section of his field. In short, he treats his one large field as two smaller fields when it comes to fertilizing and liming.

As far as taking the samples, it is pretty straightforward, but there are a few tricks. The soil should be moist enough to take a sample, but it shouldn't be damp or wet. The sample should be taken from between plants so the sample isn't mostly roots. Clear off the sample spot so that no loose organic matter from the top of the ground gets into the

sample. Take a plug from the ground a few inches across and 6 inches deep. I use a tulip planter, which gives me an adequate size sample. Slice off the top two inches of soil. Take what is left and place it into a bucket. Collect a few more samples, about one every two acres or so, and throw them into the bucket. After you've collected all your samples from the field, pulverize the sample with a clean piece of wood (don't use metal) and mix the soil thoroughly with your hands. Fill the soil sample box or bag from your agricultural extension agent and return it to the agricultural extension service to be analyzed. The office will either mail or call the report back to you (some states place it on the agricultural department's website for you to retrieve).

Typically the bottom-line results are pretty straightforward. However, the interpretive part of the report should be reviewed with the agricultural agent's help since much of the chemistry they use in this section may be a bit

Above: A cultipacker improves germination rate of your field. As it passes along, the soil is compacted, which improves the seeds' soil contact. A cultipacker-pulverizer has geared sprockets to help break up small clods of soil and improve the finish of the soil. Below: A true pulverizer can clean up tillage mistakes. If soil moisture conditions weren't optimal when you worked your field, and general roughness and abundance of dirt clods seem to be a problem, you may need to use a pulverizer after discing to improve the finish before drilling. The pulverizer will break up dirt clods and smooth out the field.

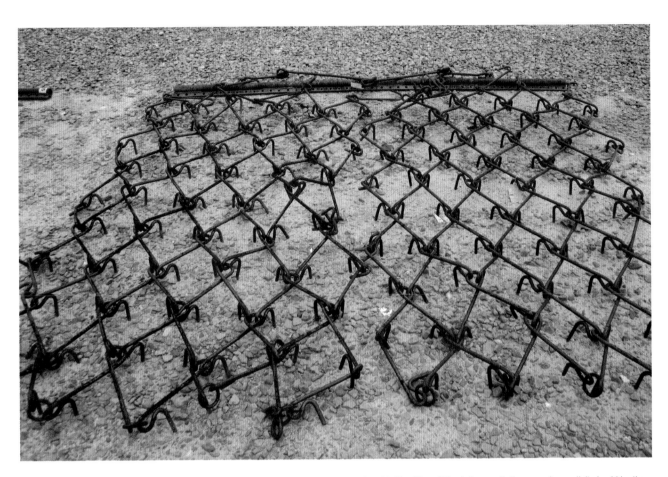

A field rake (is known by many names) will clean up field trash, such as plant material, left behind by tillage. It tends to smooth the ground as well. It should be the last step before drilling your grass seed.

overwhelming. I included one of my reports as an example. Also, if you use a custom fertilizer blending/spreading company, they can take the report and custom blend a mixture for you that will be extremely accurate and faithful to the report's recommendations. I recommend using the agent on your first test, and then talk with your fertilizer supplier on subsequent tests.

You do need to know one important thing about nitrogen. Forage and hay crops can be toxic to animals through the accumulation of nitrates in the plant. While the exact mechanism that causes hay, forage, and crops to accumulate high levels of nitrates isn't fully understood, plant stress such as drought, hail, and traffic, in addition to over-fertilization, is a significant risk factor. For this reason you should test your soil every year during periods of low rainfall. Also, never fertilize above the rate recommended in the test for this reason.

In most soils, controlling pH will, at least during the first application or two, improve the fertility of your field as much as fertilizer. In acidic soils, lime is used to control pH. In alkaline soils, an acidic fertilizer is used to balance pH, which is important because plants absolutely require that the soil they reside in have the correct pH to efficiently bring nutrients into the root system. In other words, the plants' uptake of nutrients is determined almost solely by pH. However, the pH level of your field will return to its native level over a period of a few years, so most fields, unless they have been actively and careful farmed in the recent past, will have pH levels that are incorrect for hay. Therefore, do not skimp on any recommendations made for correcting pH. I never skimp here and I put out lime every time I fertilize, no matter how little the test calls. Lime takes a while to dissolve and fully neutralize the soil. Therefore, the pH measurements will swing widely unless you stay on top of maintaining the pH of your field by adding small amounts every year instead of large amounts every other year. Your agricultural agent will help recommend proper lime rates and application strategies.

APPLYING FERTILIZER

The soil test report will tell you how much fertilizer and lime you should spread. However, that amount should be ideally split in half and be applied twice—once in late winter and once in late summer. However, it can be difficult to do depending on your weather and climate since late winter fields are often a bear to get equipment into and out of without damaging the field. On the other hand, applying all your fertilizer at once may create a nitrate problem. If your report indicates recommends quite a bit of nitrogen—more than 120 pounds per acre—consult with your agricultural extension agent for the number and timing of applications. (Growing a pasture with 10 percent clover drastically reduces the need for nitrogen and eliminates the risk of nitrate poisoning.)

You can apply lime and fertilizer with a pull spreader or a broadcast spreader. I prefer pull spreaders because I can better control the rates and applications. Also, no matter how well the broadcast spreader is designed, you are pelted with fertilizer and it ends up in the nooks and crannies of your tractor and your clothing. But I still encourage you to look into spreading services. In fact, I no longer apply my own fertilizer and lime. Typically, if you use the spreading service of the company you purchase your fertilizer and lime from, the cost is quite negligible and is much less than a spreader and way less than the value of my time spreading it. I think only on fields smaller than five or six acres should you consider spreading the fertilizer yourself.

When do you call the fertilizer distributor or fertilizer spreading company to bring out your fertilizer? Actually, first—before you plow your field in preparation for planting. Why? Easy. Fertilizer and lime will be much more beneficial to all crops, hay included, if it is incorporated into the top 6 to 8 inches of the soil. If you lime and fertilize first and then perform the tillage operations, you will put the nutrients and lime down where the newly germinating plants need them. Working these additives into the soil helps to prevent run-off of the additives. This stuff is too expensive, and too harmful to streams and lakes, to allow much run-off.

CHOOSING THE TYPE OF HAY TO PLANT

The type of hay you plant depends on local market conditions, your livestock needs, and the recommendations of your agricultural extension agent. We are too warm in this part of North Carolina for traditional grasses like bluegrass and we are too cool for grasses like Bermuda or Sudan grass. That leaves us with planting alfalfa, fescue, or orchard grass. In this area, the horse owners' market is the only one lucrative enough to sell to, and orchard grass mixed with a bit of timothy, bluegrass, and crimson clover sells well. Fescue, while once popular, has fallen out of favor. Around here you have to sell alfalfa or orchard grass. What you prefer to feed your animals is also important to consider.

Your agent will have great advice. For example, we can grow bluegrass, but it doesn't produce enough hay to justify the time, effort, and expense. However, when planted with something else, it grows better and makes an excellent addition to a grass field of another variety.

PREPARING THE FIELD

Now you are ready to till the soil. In virtually all cases, new fields of hay should be started by a full plowing of the field. This is because most hay varieties are susceptible to weed competition, and a fresh clean field plowed in the late summer gives the hay a chance to establish itself through the winter before weed competition begins. A traditional turning plow can be used behind your tractor to plow the field. If you don't own a plow, ask a neighbor if they can plow it for you for a fee/exchange.

After the field is plowed you must disc and level the ridges. A traditional tandem harrowing disc should be used with a drag log. The log should be as long as your disc is wide. The log should be heavy and about 12-18 inches in diameter. The log is then chained behind the disc with its axis perpendicular to your line of travel. This will provide additional help in leveling the tillage ridges. Depending on the soil moisture and type, you will probably need to pass over the field twice with the disc to smooth and level it to an acceptable degree. The rule of thumb is the discing should completely obscure the plowing ridges and leave very few clods of dirt larger than a softball.

The next step is to travel the field with a harrow rake (not to be confused with a disc harrow). This rake looks like a chain-link fence or wrought-iron gate with teeth sticking down. The purpose of the rake is to pick up trash (roots, stick, vegetation left over from the plowing, etc.) and provide a finer finish to the field than a disc and log can. If the tilth of the soil is already good as you begin your second discing, the log can be removed. The harrow rake is attached behind the disc.

PRE-EMERGENCE EROSION CONTROL

There are a few things you can do to prevent erosion while tilling and planting. For starters, you should plow, disc, and plant against the fall of the land. Even though the purpose of these activities is to make the soil very smooth and even, there always remains, in most soils (especially clays), a grain that follows the direction of travel of the implements. In these soils, the soil will settle initially with a very slight ridged effect (it will subside over a season or so) from these tillage operations. This ridged effect, if perpendicular to the fall of the land, actually has a noticeable effect in preventing minor erosion. Cutting several very small ter-

This three-point fertilizer distributor is widely used on small-acreage operations. The capacity (400 pounds, as the model number suggests) is right for our operations and just about any tractor can carry it. If you buy one you'll need some flexibility in application rates. The top right picture shows the adjustment mechanism. The application scale, 0 to 8 in this case, is arbitrary and doesn't actually mean anything in units. Setting it up for the correct rate can take some practice. I have practiced with sand to get the handle on getting the application rate right. You'll also need the ability to sling fertilizer to the right, left, or center, which this implement allows. The most important thing is making sure that whatever one you buy has controls that are accessible while on the tractor. You also must make sure the agitator stays clean during use. Any trash, such as leaves and sticks, will clog this up in a heartbeat.

A drop distributor like this is a tow-along that drops fertilizer and lime as it travels.

races into the hills with a single bottom plow after planting will also help.

In very steep areas or erosion-prone areas, dump large rocks in strategic areas to slow water. Remove the rocks before your first harvest and plant in the bald areas left behind by the rocks. Using bales of hay to prevent erosion is helpful, too. It is expensive, but keep an ear open for someone who may have some old hay for disposal. When you place bales in the field, be sure to place a stake through them and into the ground because heavy rains will push the bale out of the way.

PLANTING

The next step is to plant the field. I highly recommend using a seed drill, which promotes excellent germination, minimizes the wasting of seed, hides the seed from birds, and protects the seed from heavy rains that will wash it away. Since the field is plowed and most hay species only need to be planted ¼-½ inches deep, any grain drill will handle the job. Even one of those old $200 drills found at the local implement yard will work, but make sure the drill either has a grass-seed box or a setting for planting rye and/or wheat. These seeds are comparable to grass in size and any drill that

A small tandem disc is required for smoothing and leveling the field after plowing. This angle between the tandems is not adjustable on this particular model, and getting a slightly larger disc that is adjustable might be a better choice.

can plant rye or wheat can plant grass seed with very little waste. There is an implement called a cultipacker (see below) and certain makes and models of these have grass-planting attachments. These create a nice stand as well, but tend to be pricey—even older used ones. The same broadcasting implement you might use for fertilizer and lime can broadcast seed as well, but it broadcasts seed on top of the ground, which is not ideal in most soils and climates. Therefore, I recommend running a rake harrow over the ground after seeding to try to get the seed under the surface.

The rate you plant the seed is highly variable to the variety, soil, weather, irrigation strategies, and more. Below are some typical seeding rates for various cold weather varieties across the country:

Seed Type	Depth to Sow	Pounds to Sow Per Acre	Time to Sow
Alfalfa	¼–½	10–20	Spring and early fall**
Barley	1¼–2	72–96	Early fall
Birdsfoot Trefoil	1/2	5–8	Fall and spring
Bluegrass-Lawn*	1/2	20–40	Anytime
Bluegrass-Pasture	1/2	15–20	Early spring and early September
Bromegrass	1/2	8–10	Spring
Buckwheat	1–1½	50–60	Late spring
Clover, Alsike	1/2	4–6	Winter to April
Clover, Crimson	1/2	15–20	Winter to April
Clover, Ladino	1/2	½–2	Spring and fall
Clover, Mammoth	1/2	8–10	Winter to early April
Clover, Red	1/2	8–12	Winter to early April
Clover, Sweet	1/2	15–20	Mid-Winter to early spring
Clover, White Dutch	1/2	½–3	August. October, and spring
Corn Silage	1½–2	10–15	Spring
Crownvetch	½	5–15	Spring and fall
Fescue, Creeping Red	⅛–¼	3–5	Early spring
Fescue, Ky. 31	½	10–20	Early spring and early fall
Fescue, Tall	¼–½	10–16	Early fall
Green Needlegrass	¼–½	5–7	Early fall
Lespedeza	½	10–25	Mid-Winter to early spring
Millet, German	1–1½	10–20	Late May to early July
Millet, Pearl	1–1½	20–25	Late May to early July
Milo (for Grain)	1–1½	4–8	June 15 to July 1
Oats	1½–2	32–96	January 20 to March 20
Orchardgrass	¼–½	5–8	Early fall
Orchardgrass	½	10–20	Early spring and early fall
Redtop	½	5–6	August 15 to October 30 and spring
Reed Canarygrass	1½	6–8	August 15 to September 30
Rye Grass	½	2–25	April or early September
Rye, Winter	1½–2	72–112	September to October
Ryegrass, Perennial	¼–½	14–18	Early fall
Sorghums, Forage	1–1½	10–12	May to June 20
Sorghum-Sudan (for Pasture, Greenchop, Hay)	1–1½	20–30	Late spring
Soybeans (rows)	1–1½	35–40	Late spring
Sudan for Pasture, Greenchop, Hay	1–1½	10–20	Late spring
Timothy	¼–½	6–9	Early fall
Timothy	½	6–10	Early fall
Vetch, Winter Hairy	1½–2	20–30	Early fall
Wheat, Winter	1½–2	60–90	Mid fall to late fall
Wildrye, Canada	¼–½	8–10	Mid fall

* provided for comparison
** early fall includes the last four weeks of summer for some areas of the country

If rocks are a problem in your area, a rock picker is recommended. After plowing and discing, travel the field with this rock picker. It will make short work of field cleanup.

After planting, you should travel the field with a culti-packer/pulverizer. This device looks like a set of iron rollers and gears that turn on an axle as you pull it along. The purpose of this device is to improve the seeds' contact with the soil and pulverize the clods of dirt that the disc harrow and rake harrow missed. While the compaction isn't vigorous enough to provide a crust on the top of the soil, it improves the structure enough to prevent small rains from washing the newly plowed, disced, and planted earth.

After planting, it's time to turn the controls over to Mother Nature. Let the sun shine and the hay germinate and grow. In a few months you should have a full, but short stand of grass that will continue to grow and strengthen a root system all winter long.

However, while the grass is being established, you run the risk of soil erosion. One of the risks associated with planting in the late summer in the southeastern United States is water erosion caused by the heavy rains from tropical storms and hurricanes that occur then. Erosion can be quite a problem unless your field is very flat. You do have

to consider how you will handle erosion until the grass is high enough to prevent erosion. Hopefully you won't need these precautions, but there is nothing worse than staring at a 2-foot deep gulley slicing through your field that was caused by an unexpected heavy rain.

WEED CONTROL

After the stand establishes itself, there is very little care the field requires other than regular harvesting and the yearly application of fertilizer and/or additives to correct soil pH. Weed control is a task that may need to be added to your field maintenance checklist. In most areas of the country, weed control is not an issue because a strong, healthy stand of hay grass that is regularly harvested and maintained will out-compete virtually all weeds. Of course, there is no such thing as a weed-free hayfield, but a handful of weeds per acre is to be expected and can be tolerated. However, drought, missed maintenance, and other factors may give weeds a leg up on the grass/forage plants and may need to be actively controlled.

Any broadleaf weed killer can be used; there are about three different chemicals in use. The most widely used, and the only one I can recommend for virtually all areas, is a chemical with the chemical shorthand of 2-4D. There are many trade names for this chemical; Weed-B-Gone is one well-known from the private lawn market. This chemical typically shows good results. Your farm supply center can help with specific recommendations as to the rate and timing of application for your area. Other chemicals are not as widely used, but are popular in certain areas. Other chemicals are popular for certain crops. With alfalfa, Pronamide is used quite a bit both for early use during seeding/emergence and for established fields.

It is important to remember that 2-4D and most all other broadleaf weed control products will kill clover and some other types of forage/hay crops. If you are having trouble with weeds, I highly recommend contacting your agricultural extension agency. Not only can the agent point you to the right chemical for the weeds you are having trouble with, but more important, you can get some tips and tricks on applying it safely.

Since you only establish a hay field once every so often (2 to 3 years for alfalfa, 5 to 10 years for other varieties), tillage and planting equipment is better rented or borrowed. You can also hire out the work or buy the equipment yourself. Tillage equipment typically isn't horribly expensive.

PLOWS

There are two types of plows: moldboard and disc. Disc plows do an excellent job and are widely used. They also are commonly found on larger plows pulled by large tractors. Smaller plows are typically moldboard plows and are come in two varieties: three-point hitch or trailer plow. Three-point hitch plows hook to the three-point hitch of your tractor, and the draft-sensing function of your hydraulic system helps maintain consistent plowing depth. Trailer plows hook to the drawbar of your tractor and use a series of hydraulic cylinders or levers and wheels to maintain depth. Trailer plows produce better fields but are hard to find and are less convenient to use. Three-point hitch plows are more common and work well for our purpose.

DISCS

After plowing, the ridges must be broken up and leveled. A disc harrow does this job perfectly, especially when outfitted with weights and a drag fastened behind it. While these discs come in a multitude of sizes and configurations, the only configuration we are interested in is the tandem axle, adjustable disc harrow. This disc has two axles of discs (actually, two lines of discs with four axles) that can be moved in relation to each (hence the name adjustable). The double line of discs is more efficient in terms of leveling and the adjustable aspect provides aggressive/less aggressive cut

When searching for a drill, it would be great to find one that has a grass seed planter attachment. This makes obtaining correct planting rates very easy. The grass seed planter is the small set of boxes on the front.

You may find that a front loader is very handy for loading and unloading implements, carrying bags of fertilizer, and moving round bales (using a baler spear), in addition to other farm chores. Fortunately front loader attachments for many old tractor models are still available new or on the used-equipment market. This one for the Ford 600 series is from the late 1950s and is in great shape.

into the dirt, which affects how completely the plowing ridges are knocked down.

SPRING TOOTH HARROW/RAKE HARROW

This implement is known by several names and looks like a chain link fence with teeth pointing down. The harrow provides the final leveling and will clean trash out of the field. If you broadcast seed, you should rake again after broadcasting to improve the soil contact with the seed. There are several brands and sizes of harrows, but any will work well. Since they don't provide much of a load on a tractor, I recommend getting the largest one you can turn with and afford.

SPREADERS

There are two type of spreaders: drop and broadcast. Broadcast spreaders look like funnels and sling the seed/fertilizer out with a rotating disc found at the bottom of the funnel. Broadcast spreaders are usually attached to the tractor via three-point hitch, but there are draft-style broadcast spreaders, too. Drop-style spreaders are metal bins on wheels with a small slit along the entire bottom of the bin. A gate in this slit is adjustable to allow seed and fertilizer to drop out of the bottom at a consistent and controlled rate. This rolls along on wheels and attaches to the drawbar of your tractor and is never a three-point implement.

DRILLS

If you do not broadcast the seed, you will have to drill the seed. The word drill is misnomer, at least in modern usage, because the implement doesn't make small holes. It cuts slits in the ground as it moves along and drops seeds in the slit. There are two varieties available. One comes with a culti-packer/pulverizer. Brillion seeders are the most notable and do a very nice job of seeding a pasture. The second type of drill is the average grain drill. It may or may not come with a special grass-seeding attachment. If you choose an ordinary grain drill, make sure it has small seed capabilities.

CULTIPACKER/PULVERIZER

After planting, you will need to lightly pack the top 2 or 3 inches of soil because the grass seed requires good soil contact to germinate and be anchored as it first emerges. To do this, you will use a cultipacker. Optionally, this will have a several discs that look like gears that pulverize the clods of dirt while packing the soil. You only need to travel the field once with this implement. Any more and you may pack the soil too much and create a situation where rain will run off and cause erosion. Be sure to travel perpendicular to the fall of the land when using this implement.

OTHER IMPLEMENTS YOU'LL NEED

There are several implements you will find handy after the field is established. Every year heavy soils should be plugged to improve soil oxygen and compaction. Having a plugger is handy, but you can also rent one. A subsoiler is also used to improve deep soil aeration and compaction. Either implement works well. A small rotary brush cutter or finish mower is very handy for keeping the edges and corners of the field that the haying mower can't reach. Carry-alls (a three-point box for carrying supplies and tools) and post hole diggers round out the implements you may find handy as you raise and harvest hay.

While the work is full and intense, the decisions, implements, and costs are nerve-racking and exhausting, and the outcome is unsure, planting a field is rewarding. Watching it come up and fill out is a really enjoyable experience that instills a sense of pride. Just make sure you think ahead, dream and plan completely, enlist the full support of your agricultural extension agent, and talk with others who raise hay. If you do these things, the odds of success are very high and the quality of the hay and your field will reflect it.

TIPS AND TRICKS

In very few cases agricultural extension services recommend no-till drilling or other methods such as scarification with broadcast. These are used in arid areas and areas with unworkable soils.

HOW I SET UP MY HAY FIELD

Writing about starting a hay field is like telling someone how to plan for and prepare for a career or some other fairly complex long-term project. It is nearly impossible to do well without either over-simplifying it or going into too much detail. I think it would be beneficial for you to hear how a field is established, so I'll outline how I set up my field.

When I decided to get into hay production, I was already somewhat familiar with the hay market since my wife had a horse at the time. I didn't really know anything more about hay than how to buy it, with the exception of what I learned being a paid hand loading hay a few times as a teenager. When I made the decision to be a haymaker, I started reading and researching on the Internet. When I mentioned my new project to my representative at the local farm credit cooperative, he provided the first and one of the best pieces of advice I received. He said, "You talk too much like a horseman, not a haymaker. Go talk to some local hay producers and do some reading." I followed his advice and very quickly learned that books were scarce and that being

While not necessary, a box scraper is handy to have as it does a great job of grading off the top of high spots and carrying the dirt to low spots. I use this where necessary after discing. It also does a great job on other farm work, such as final grading on farm roads and more.

When plowing a field, a set of disc plows is tough to beat. An moldboard plow does well, too. Field plowing can be hired out for less money than you can do it yourself, especially because it is something you do only once or twice a decade for a hay field that isn't rotated for other crops. However, your hay tractor and a moldboard plow will do a great job if you take your time and be careful.

and was pivotal in preventing me from making some mistakes I would have regretted. The agent helped me find buyers for my hay, pointed me in the direction of used equipment, and made solid agricultural recommendations that always served me well. The extension agency definitely does more than help with soil samples.

The biggest mistake the agent helped me avoid was hay variety. I was dead set on raising fescue for many reasons, but the agent recommended an orchard grass mix or alfalfa. It was explained to me that fescue was slowly losing popularity among the target market (horsemen) in favor of orchard grass. Since I entered into production at a good time to take advantage of this market shift, he strongly recommended it. He also recommended alfalfa because it is very profitable, always sells well, and is always great hay for livestock.

The agent then explained some of the care and other issues of the different varieties. In North Carolina we must spray pesticides on alfalfa to eliminate the alfalfa weevil. Pieces of information like this helped narrow the list of possible varieties down to just fescue and orchard grass since I didn't want to get involved with pesticide handling and application.

Then the agent asked, "Will you be adding red clover?" I didn't realize that in this area it is very common to sow your hay crop with 10-15 percent red clover. As a legume (a plant that through nodules in the root system fixes nitrogen in the soil), clover will drastically reduce the need for nitrogen fertilizer application. Clover also germinates and establishes much more quickly than the grass, and therefore acts as a nurse crop (a crop that helps another with shade and erosion control) to the grass seedlings. Also, clover is an excellent hay crop with very high protein content.

This all sounded great, but then the agent explained that there is one drawback to clover. It has a certain type of fungus on it called black patch. During most times of year, your hay cuttings will contain this fungus in low enough amounts that it is of no consequence, but midsummer cuttings after a wet winter and spring will have clover with a significant amount of fungus. While it doesn't affect palatability, nutrition, or appearance to any extent, it will create a harmless, though unsightly, case of slobbering in horses. This is unacceptable to some stables, especially those that hand rein horses a lot and

a hay buyer is not as helpful as one might think when making the transition to hay producer. The main reason is that hay buyers harbor some old wives' tales about hay that is detrimental to quality hay production. I immediately learned that gathering information could be tough (I hope this book solves that) and that every new haymaker should try to talk to other producers, keep an open mind, and realize you might have a lot to learn, even if you have been around livestock and hay.

Along the way I learned some other valuable tips. The local agricultural extension agent is a tremendous resource

After all the plowing, discing, raking, planting fertilizer, time, sun, and water, this is the final result: a happy, straight row of hay. Some species of hay will fill in between the rows and create a carpet of grass. Other species, such as this orchard grass, will remain bunchy and keep the row structure forever. Therefore, plant against the fall line if you plant bunchy grass species.

stables that specialize in rides for children and physically handicapped persons.

I then talked with the agent about fescue and learned that I needed to consider growing something called an entophyte-free variety. An entophytic relationship is when one organism lives inside another for mutual benefit. It turns out that a fungus exists in many grasses, particularly fescues, and has detrimental effects to broodmares. This fungus can lead to a toughening of the birth sack and lactation difficulties. He explained that anyone with brood

mares will not buy my hay if it wasn't from a seed certified to be entophyte-free. On top of that, my field wouldn't remain entophyte-free over time if any of these funguses were present in other fields or lawns nearby.

I walked away from my conversation with the agricultural extension agent realizing why this country established land grant universities and agricultural extension services and why people go to college for this work. I was absolutely convinced after my conversation that consulting the agent was just not helpful, it was instrumental in preventing huge

GPS SURVEY

Hilly fields have one problem: The surveyed acreage will be inaccurate. All surveys are based on flat pieces of ground and can only measure your border and extrapolate your acreage from those borders. If your land is hilly the survey will be wrong. How wrong it is will vary, but usually the field has to be at least 10 acres before the inaccuracies start to matter. If your field rises and falls a lot, the more likely your survey is wrong. The benefit of getting a more accurate survey is to allow you to purchase and use exactly the correct amount of fertilizer, lime, and seed.

One way to get a more accurate acreage value is to see if the fertilizer spreader service you hire can do it for you. Otherwise, if you are comfortable with technology, a simple handheld global positioning system (GPS) unit and a couple of free software programs is all you need. The idea of a GPS survey is to take a lot of waypoints (latitude, longitude, and elevation of a location) and store them in the GPS unit. When I plotted mine, I walked the field and took waypoints (a total of 250) as I went. This number seemed to be sufficient, but more data is always better.

After you've collected the points, download them to a file on your computer. (Your GPS should have included software to download to a computer.) This process of downloading the data to the computer will store the points in a file using the National Marine Electronics Association (NMEA) format. This is a standard format almost all map/cartography based software and devices like GPS units understand. After the data is in the computer, download freeware acreage calculators that will read this NMEA standard file. One I have used in the past is the GPS Acreage Calculator. It is free and readily available on the Internet. Simply search the Internet using "NMEA GPS ACREAGE" with your favorite search engine and the latest version of the most popular GPS acreage calculators will pop right up. These pieces of software will read the file, look at the way points in your file, and calculate the acreage of the field using mathematical techniques that take the three-dimensional characteristics of waypoints into consideration.

On my field, the legal survey said 10.1 acres. However, at least an acre was wooded so the field, which is hilly and changes elevation more than 40 feet on 2 axes, was no more 9.1 to 9.2 acres according to the legal survey. However, the GPS survey of the cleared portion field was 10.1 acres. That inaccuracy is 10 percent (10.1 actual – 9.1 survey = 1 acre difference between "real"' and "surveyed") and is enough of an inaccuracy that lime and fertilizer rates would have been underestimated for my property. I was glad I took a few extra hours and did the GPS survey.

mistakes on my part. The office had a lot of literature and field-preparation techniques to help me get started and gave me a free soil sampling kit to help me collect my soil samples. These samples provided me with precise lime and fertilizer requirements. Armed with the handouts and some additional Internet research, I decided that an orchard grass mixed with other grasses and clover was what I wanted to sow. The orchard grass ensured salability, the clover helped reduce fertilizer needs and created a higher protein hay, and this particular mix had the least maintenance and diet and health concerns for livestock.

After you decide on the hay varieties you want to plant, you will need to go to the local farm supply co-op and ask what seeds are available. My local cooperative sells a mix it calls Pasture Maker that is 60 percent orchard grass and 30 percent timothy and bluegrass. According to my research I wanted more orchard grass than 60 percent, so I decided that when it was time to plant, I would buy pure orchard grass seed to mix in with the Pasture Maker to raise the orchard grass percentage. I then added clover to the mix and I had my seed mix for planting.

My greatest lesson in all of this was that ignorance is a tough tree to cut and being willing to learn is the only sharp saw around. While in theory I suppose research and learning is subject to the same law of diminishing returns as everything else, in practice there really is no such thing as learning too much. I learned a lot during this initial research and the understanding I now have leaves me amazed that I assumed that my experiences in hay buying and bale loading qualified me to know how to produce hay. It also makes me realize how close I came to making some bad decisions that would have haunted me until the field paid for itself and could be replanted in three to five years at the least.

CHAPTER 3
FIELD MANAGEMENT AND MAINTENANCE

After planting, there is nothing to do but harvest, right? No. It isn't that easy. Studies show that fields that are actively maintained, fertilized, and over-seeded yearly show productivity gains that outstrip the input used in the maintenance. Of course, this shouldn't surprise any of us who have tried to grow a decent lawn. You just can't plant it and leave it alone. It takes considerable work to maintain the field. While the type and scope of work differs, that same commitment to yearly and seasonal maintenance is required. Of course, there are also improvements that may need to be made. Erosion problems create the need to change or sculpt the land in a new way, fencing repair may be required, and there are more improvements that may pop up from time to time. None of the maintenance is particularly difficult or time consuming, but it does mean you won't be able to plant it and forget it.

FENCING

Field fencing is often required by many haymakers because pastures are occasionally used for grazing or grazing pastures, and haymaking pastures are sometimes rotated. There are many types of fencing, so I'll go over a few of them and make a few recommendations and outline some issues you'll need to consider. There are only a few issues during the design phase of a fence that affect haymaking. If you are going to rotate grazing pastures and hay pastures, you will more than likely need to cross fence. Establishing a hay pasture usually results in thick, highly nutritious forage for animals. Animals such as horses that can actually become quite sick from over-eating (foundering) need to be restrained on hay pastures, especially during heavy growth periods in spring and fall. Taking a hay field and sectioning it off with cross fencing into small areas that horses can be moved to every other day or so results in a stronger pasture and the prevention of bloating, colic, nitrate poisoning, and foundering. However, cross fencing makes harvesting the hay very difficult, so the design should include temporary cross fencing. Usually a double electric wire on foot-set poles will do the trick.

Two other issues to consider during design are the type of fencing material and the shape of the fenced area. Unless there is a need (i.e., keeping out smaller predators such as

NCDA Agronomic Division	4300 Reedy Creek Road	Raleigh, NC 27607-6465	(919) 733-2655		Report No: 01519

Soil Test Report

Grower: **Yost, Spencer**   Copies to:

SERVING N.C. CITIZENS FOR OVER 50 YEARS

7/29/04   Forsyth County

C -- 12

Agronomist Comments:

Field Information		Applied Lime			Recommendations									
Sample No.	Last Crop	Mo	Yr	T/A	Crop or Year	Lime	N	P2O5	K2O	Mg	Cu	Zn	B	Mn	See Note
SUPST	No Crop				1st Crop: Fes/OG/Tim,M	.3T	120-200	30-50	0-20	0	0	0		0	12
					2nd Crop: Fes/OG/Tim,M	0	120-200	30-50	0-20	0	0	0		0	12

Test Results

Soil Class	HM%	W/V	CEC	BS%	Ac	pH	P-I	K-I	Ca%	Mg%	Mn-I	Mn-AI (1)	Mn-AI (2)	Zn-I	Zn-AI	Cu-I	S-I	SS-I	NO3-N	NH4-N	Na
MIN	0.86	1.24	7.3	85.0	1.1	5.8	41	76	53.0	27.0	157	111	111	183	183	218	58				0.1

This is a copy of my grower's soil test that was taken during the last field maintenance. A soil test is something you should do every year prior to annual field maintenance. This report is issued by the state of North Carolina, but should be pretty similar to what any grower would receive in any state. The recommendations are a bit tricky if you aren't used to calculating fertilizer ratios, but it isn't too hard and they usually come with a handout that will show you how to do it. Your fertilizer supplier will custom-blend your fertilizer based on the report, so calculating the ratios yourself usually isn't necessary. Even if you are applying the recommendations yourself, the farm supply store where you buy your fertilizer and lime will be able to help.

Begin your yearly field maintenance and management after your summer cutting. You will want the hay crop shorn close so the fertilizer and lime makes it into the soil and the multiple passes you'll make with the equipment will do less damage to the leaves of the hay crop. This picture shows the right height, but the little flush of growth indicates the hay has recovered from the stress of harvesting.

coyotes), try not to use fencing material that goes all the way to the ground. Your equipment will snag and catch on it and cause damage to the fence and/or equipment. Consider high-tensile fence instead.

The other issue is the layout of the fence. Design your fence with rounded, not sharp, corners. The hay in sharp corners can not be mown with hay harvesting equipment. Also, sharp corners tend to trap animals being pursued by another animal (i.e., one horse bullying another horse). Granted, it isn't the end of the world if you do design using sharp corners, but leave enough room between the field and the fence so you don't damage the fence while harvesting the hay.

If at all possible, build the fence after planting is done and the pasture is established. The tillage equipment used for planting is usually large and difficult to maneuver. Operating this type of equipment near a fence is a bit aggravating and the fence is likely to be damaged. In short, as you are designing and building a fence, remember that you will have to operate clumsy large, heavy equipment near any fence you build and make provisions in the design and selection of material accordingly.

SOIL EROSION/FIELD TOPOLOGY CORRECTION

Soil erosion is a constant problem in many areas of the country. Even if water erosion is not much of an issue in your area of the country, the practices used to prevent it are many of the same ones used to maximize soil moisture retention. For example, making many small terraces or cuts perpendicular to the fall of the land is very popular in the south for preventing erosion. In the more arid parts of the inner-mountain west these same small terraces are used to collect water and snow melt. The water soaks through these open terrace cuts and the water is stored in the soil and is available for the plants of the next growing season.

Wind erosion is less of a problem in hay fields than the fields of other crops, although hay crops should be drilled perpendicular to prevailing winds so the winds can't "run the rows." During field establishment or right after harvesting, the wind has access to the soil, and at those times the wind can cause excessive soil moisture loss. If you live in a dry, windy area, conserve moisture by making sure you cut your hay high when harvesting. The extra height of the grass will help protect the soil and will minimize soil

moisture loss. Your agricultural extension agent can advise you on specific erosion issues that might exist in your area.

Occasionally you will have to change the lay of the land to create a successful hay field. This occurs for many reasons, though the two most common are soil drainage and areas where the lay of the land doesn't lend itself to safe, easy navigation with the hay equipment. An example of the latter would be a steep cut or bank near the edge of a flood plain. To correct drainage, you will need middle busting plows to create drainage catches through the field to direct water away from the highest areas and to drain low areas. If the amount of water that drains is reasonable, then a deep cut with a middle busting plow (a plow that throws dirt in both directions) will be all you need to correct the drainage. If the amount of water is more significant, you may have to hire a grading contractor to perform the work.

Cutting and leveling high spots and banks and general leveling can be done with a large offset-disc plow. A few passes with that and the bank or high spot will no longer exist. If you don't have access to a large tractor and this type

Part of field management is aerating your hay field, and here we are getting ready to use a "plugger" to do just that. In clay soils that are dry from the summer heat, some extra weight (shown here in the form of water) is needed for adequate penetration.

As a safety precaution, many implements that are top-heavy have stands, such as the four stands this plugger has (two are visible in this picture). Stands prevent rollover and instability when the tanks are full. We hitched up the implement and raised the stands so you could see them. There is also a higher traveling position for the stands that is used while operating the plugger.

of disc plow, patience and many trips with a box blade should take care of the worst of the problem. Just remember to be careful as you work on steep areas. Most tractors have a rollover hazard and working on banks with box blades and scrape blades is a common way turnovers occur.

PESTICIDES/HERBICIDES

Before we get started with this section, I need to mention a few things about safety. These chemicals are dangerous. Their long-term effects are not as widely understood as the manufacturers would have you believe; and improper usage, mixing, or application will create a public health hazard. Under no circumstances should you apply any pesticide or herbicide that you don't fully understand. Under no circumstances should you apply these chemicals without giving some thought as to how you will respond to a spill or how will you react to toxic exposure. While most of the chemicals you will apply rarely, if ever, create a need for an emergency hazmat response, it is advisable to have help nearby or have a cell phone handy while handling the chemicals.

The plugger is ready to travel. The weight of the water, which is about 500 pounds, demands that the implement is perfectly level after mounting to minimize stress to the hitch linkage. Notice the use of the three-point-hitch conversion bar. Any implement that doesn't use the PTO will benefit from a three-point hitch because it allows you to hitch to just a drawbar and make tight turns.

There are also a couple of general notes to be made regarding application. The most important aspect of the whole process is getting the application rate correct. Not putting enough chemical down may be a waste of time and will unnecessarily create risk without any benefit. However, applying too much creates additional risks unnecessarily and wastes money. Another important point is that application rates are tough to get exactly right since the ground speed of the tractor, pump output, and other factors influence how much chemical goes down per acre. It is helpful to fill your tank with plain water and apply the water as if you were applying a pesticide or herbicide. This will give you the practice you need to apply your chemicals in exactly the right amounts. Another important note is that chemicals are nearly universally applied in the early morning when the air is still and the air is still sinking from the radiational cooling from the night before. Respirators are always a good idea when applying chemicals, and the material data safety sheets should be consulted to determine what safety precautions must be taken during mixing and application. By law, your chemical supplier must give you one of these MDS sheets.

Here is an example of setting up an application of 2-4D, a broadleaf control chemical sold under several brand names that is used by many hay farmers. First, determine how much of the chemical per acre you will need. Typically,

The working end of the plugger is shown here. The plugger tines are slightly curved to ease insertion and removal from the ground. The tine is hollow and the plug falls out from side of the tine as the tine moves upward to the sideways position. In this photo you can see the lime on the plug.

farmers use the granular form of 2-4D, and the amount used per acre varies quite a bit—from less than a pound to nearly four pounds depending on several factors. Once again, your agricultural extension agent can be of help here. For our example, we will assume two pounds for acre. Typically, you dissolve 10 pounds of 2-4D in 100 gallons of water. This provides a concentration that is workable in terms of spraying time and ground speed. Since we need to

This is what the field looks like after it has been aerated with a plugger. The first is a close-up of the plugs, and the second gives you a general idea of the number and density of the holes. This will dramatically improve aeration. If you overseed by broadcasting, plugging is almost mandatory to provide the room necessary to nurture germination.

apply two pounds per acre, we will adjust pump output and ground speed so that the entire tank is emptied in five acres.

As far as the equipment is concerned, agricultural sprayers are fairly inexpensive and simple. Agricultural spraying rigs are typically HDPE plastic tanks mounted on one or more axles. A pump may or may not be mounted on the rig. Often it is assumed that a PTO-mounted pump is already owned by the farmer and the spray rig has hoses to connect to the PTO pump. Sometimes the spray rig has its own pump with a PTO shaft or it will have an electric pump

designed to run off the tractor's 12-volt electrical system.

Mixing the chemicals is rather easy because these tanks are graduated in gallons and it is easy to see how much water is in the tank. Add the chemical in the proper ratio and the agitation caused by travel to the field will typically mix the ingredients in short order. If not, a mixing paddle powered by a drill will help. However, you can usually count on the label on the chemical mentioning if the chemical requires extensive or power mixing. Making your own spray rig is easy if you are so inclined, as almost all parts are

When overseeding with a drill, be sure to set your drill down to the slowest possible seeding rate. You can also stuff a rag or something similar in one or more of the seed discharge holes to further slow the seeding rate.

widely available at any farm supply store and on the Internet. The only tough part may be coming up with a tank that is large enough for the job.

Pesticides are a completely different matter, safetywise. Most are much more toxic to humans than herbicides and require special care in handling and application. Some even require licenses or special education certificates before you can buy them. On top of that, many are very persistent (last more than a few weeks) in the environment and can cause serious problems if mishandled. In short, I highly recommend having a professional agricultural spraying contractor spray pesticides on your field. One of the silver linings in pesticide use is that there are usually several pesticides that will get rid of a particular pest. You can often choose one that may be a bit more problematic to apply in terms of cost, convenience, and application rate, but would be safer to handle and apply.

Just remember that while some chemicals are perfectly safe for you to apply, others aren't. Also, remember to rotate brands and types of chemicals from year to year to help prevent giving pests and weeds a chance to develop a resistance to the chemicals.

FERTILIZING

One of the yearly tasks in the maintenance of hay fields is fertilization and liming. (The actual task of fertilizing is covered in Chapter 2.) You will need to take a soil test at least every other year and apply that recommendation from the soil test (which will be stated in pounds per year) each year until the next test provides newer information. Applying the whole amount in the fall is up to you. You

Make sure you drag chains behind your seed drill, even when overseeding. Anything you can do to improve soil contact with the seed is not a wasted effort.

can apply the entire recommendation in the fall or you can apply one half of the recommended amount in the fall and the other half in late winter. Remember to also include lime (if needed) in this pass, along with the fertilizer. In most soils, a yearly application of lime in small amounts is required.

Either way, fall can be busy because you will have to take another soil test if needed and get at least half the fertilizer and lime applied. Heavier soils will also have to be aerated. A perfect example is the heavy clay soil in the south. Yearly plugging is required and many hay farmers will subsoil as well. Plugging is taking small finger-sized plugs of earth from the ground and laying them on top of

When drilling as a method of overseeding, you must travel very slowly, and ideally in damp soil, to make sure your drill discs penetrate the soil crust. Here the depth and tension levers of the drill are standing nearly completely upright, which indicate a strong downward pressure on the drill discs. Coupled with slow travel there is about ½-inch soil penetration, as the second picture shows, even through roots and plants crowns.

the ground, leaving millions of small holes across your hay field. There are advantages to these holes mentioned below, in addition to the improvement they make with oxygenation of the upper 6 inches of soil. A subsoil is a larger shank that looks a lot like a chisel plow and is designed to run 12-18 inches deep and provides deep aeration and some moisture-retention capabilities to the soil.

OVERSEEDING

At least every other year, if not every year, you should renew your hay field by back planting or overseeding. This means going over your field in the fall with a drill or broadcast seed spreader and planting more seed. This is usually best done shortly after and coordinated with the fall application of fertilizer. Seed simply scattered on the ground has a terrible

When you are using the seed drill to plant, you also need to watch the seed fall through the drill tubes. As the seed agitator in the seed box rotates, the seeds fall through the seed tube. Grass seed is notorious for clogging, and wind takes the seed away from the tube, so keep a close eye on this.

Keep a close eye on the drill and its operation. This is an operator platform, and having a helper keep tension on the levers and keep the seed bins full really helps.

germination rate, so I always make sure to plug or aerate the soil after fertilizing and before overseeding. That way fertilizer and seed will be trapped in the plug holes and a rainfall or two will move dirt back into the plug hole along with the seed and fertilizer to give the seed a reservoir of fertilizer, dirt, and water. Germination rates for this type of planting are much higher than simple broadcasting of the seed.

If you are growing a mix of hay varieties, then you should consider overseeding with a mix that emphasizes a variety that tends to be weakest in the field so that the field has a solid number of plants in all varieties. For example, clover tends to die out after a few years, and in my field, the timothy and bluegrass are usually dominated by the orchard grass. Therefore, when I overseed, I always include some

One of the benefits of overseeding with a drill is that you can concentrate seed applications in the areas that need it. Here in the lower end of the field, erosion made the soil poor. The following year fertilizer application was concentrated on this area and it was overseeded with a mixture of regular seed mix, extra clover seed, and oat seed. I double-drilled only in the drainage area as the first photo shows, then I drilled that same seed mixture against the fall line a few times to help slow down rainwater movement.

Shown here is one of the drill discs penetrating the ground at first setup. Enough pressure is applied to get about ½- to 1-inch penetration. As you move, the disc will ride up a bit to give you the ¼- to ½-inch penetration that you need for successful germination.

Watch for clogged discharge holes during the slow seeding rates of overseeding. The seed will pack and clog, especially if there is any debris in the seed. Pictured here is a "drain hole" in the seed that indicates the seed is feeding properly.

During maintenance, don't forget to repair bare spots. They shouldn't require any special care other than you want to make sure you improve germination rate as much as possible. While packing and rolling seed isn't recommended during overseeding, you will probably need to use a cultipacker/pulverizer to pack seed in the bare spots.

clover and increase the percentage of timothy and bluegrass by 20 percent or so. As part of overseeding and fall maintenance, now is also the time to repair erosion and drill a nurse crop, such as oats or vetch, into those repair areas to help prevent further erosion during winter rains.

CROP CHANGES

Over time, you will find that no matter how much you fertilize and overseed each fall, the field still looks weak and sickly. The change will be subtle. Some plants in the field are happy and reasonably productive, but the hay plants are becoming sparse, open spots are developing, weeds are becoming more of a problem, and yields are slipping.

What's going on? Hay is a crop just like any other crop and will deplete the soil and need to be rotated from time to time. The only exception is a hay variety that is native and happy in your area. For example, bluegrass in Kentucky probably never has to be rotated.

Your rotation plans should be discussed with your agricultural extension agent. Here in my area, orchard grass is plowed under and legume crops are planted for a few years. For example, orchard grass fields give way after three to five years to either a year of fallow (not being used at all) or two years of alfalfa, which in turn gives way to one year of fallow. Then the field is planted with early soybeans and after an early harvest it is planted back in orchard grass in September and the cycle starts anew. Crop rotation also allows you to adapt and change to new varieties, new trends in the marketplace, and new needs that you have for your animals if you are raising hay for yourself. It also keeps you thinking about future development plans and scheduling changes, which means you are running the farm and not letting the farm run you.

OTHER MAINTENANCE

Some of the other maintenance that is performed yearly on the hay fields are nonharvested mowings, edge trimming, and rock/trash pickup. The nonharvested mowings are done for several reasons. A common reason is a drought that has robbed production to the point that harvesting it isn't worth the effort. Another is dead ends on the leaves. A hay

Field maintenance is one of my favorite parts of haymaking. It is usually done in the late summer when a little bit of fall is in the air, the quality of daylight improves brightening colors, and the flush of growth from the seeding and fertilizing provides a stunning field. This picture shows how strong and healthy a hay field can and should be if you follow these maintenance procedures.

This three-point sprayer is used for spraying herbicides and pesticides over small acreages. It has a standard PTO-driven pump and the working width is about 8–10 feet. The arms fold upright for traveling. For areas that don't need complete coverage, these sprayers are easily modified to include a spot sprayer for spraying individual weeds or around buildings, fences, etc.

This is a close-up of a PTO-driven pump. It is mounted to the PTO shaft and tethered to the tractor through the use of a chain or strap to prevent the pump from turning. It is tempting to use the pump for other purposes, but once it has been used for pesticides and herbicides, it should never be used for irrigation, livestock watering, etc.

If you raise livestock, you have a ready source of fertilizer in the form of manure. It is a great additive to pastures, and a manure spreader, pictured above and left, makes it possible. A set of chain-driven bars slowly feeds the load of manure to the back of the machine where paddles break it up and spread it on the field. Rust and wood rot is a real problem with these machines so keeping it clean between uses will be important.

field that goes into the winter with longer-than-average leaves will emerge in the spring with leaves that have partial or full leaf death on the plants. In this circumstance you will want to mow the hay back to the ground very early in the spring before growth has even really flushed out to remove the dead portions and leaves.

Late winter is also the time to remove rocks and other debris that has worked its way up through the surface of the soils by the winter's freeze/thaw cycle. About twice a year you should mow and clear the edges of the fields. Trees may

encroach on your fields. Keeping seedlings cleared back helps keep the hay equipment from being damaged by seedlings as you work the edge. Spraying Round-up™ under fences in the spring and late summer will minimize the weed eating and mowing you will have to do around the fences.

In fact the list goes on because maintenance never ends. To bend an old saw, "A farmer's work is never done." It's true. There are always fields to clean; edge rows and forest edges to keep clean; equipment to maintain, repair, and paint; and fields to plant and harvest.

CHAPTER 4
HARVESTING

Why are there so many clichés about hay harvesting? Sayings like "Make hay while the sun shines" and "It's not done until it's in the mow" litter the American vernacular. These all reflect some aspect of hay harvest but share one thing in common: They all convey that sense of work load, intensity, and accomplishment inherent in hay harvesting.

No other crop requires as much skill and technique to harvest as hay does. You must manage a half a dozen (or more) pieces of equipment, fuel, help, and weather and bring them all together like pieces in a puzzle. Not only that, the quality of your product depends directly on how successful you were in managing these variables. For example, if you mismanage your help, your haymaking becomes overwhelming or unprofitable, depending on which way you err. Mismanage around the weather and your hay will get rained on or over-cured. Mismanage your harvest timing

and the quality of your hay will suffer. Mismanage equipment and you'll be stuck with broken-down equipment in the field. Trust me, I use these examples on purpose because I have made all of these mistakes.

Contrast this with corn harvesting, which requires operating a combine and trucking the corn to a dryer or silo. That is pretty much it. The only prerequisite is the field must be firm enough for the equipment and it shouldn't be raining. I make this comparison because I want you to realize that small-scale haymaking will surprise you in terms of the skills you must possess or learn that are outside of typical farming skills. These skills include time and asset management, work flow, and forecasting (both process and weather). In short, hay harvesting is taxing, time-consuming, strenuous, and fairly complicated as far as farming goes. It is also rewarding, healthy (if you observe the safety rules), and enjoyable.

New Holland equipment is well-liked and respected in most all areas of the country. This model 488 mower-conditioner is an excellent choice for small-acreage haymaking operations, although it will be priced a little on the high side for a used implement compared with some other suitable models.

Left: This picture shows the travel stay. Many mower-conditioners have a mechanical device that holds the weight of the mower-conditioner when it is traveling or being stored so the weight isn't resting on the hydraulic ram. The travel stay must be moved/removed before you operate the hydraulic ram or you will risk damaging the conditioner or hydraulic system. Many hay farmers have done this, including myself. The hydraulic hose is also bound to the ram support to prevent pinching and rubbing by the tongue of the implement. Small preventative measures such as this go a long way in helping prevent breakdowns. Right: The mower-conditioner crimping rollers must have their tension adjusted from time to time. This turn-crank is a common method of adjustment, although simple nuts on the end of spring anchors are used on some makes and models.

It is usually impossible to harvest at the perfect time. This is a example of a hay field that has matured too far before harvesting. About 70 percent of the grass has fully headed. Unfortunately, in years with a wet spring, this will be hard to avoid.

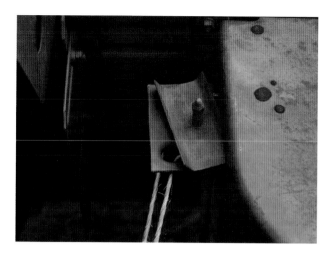

Twine tension is very important to knot production. Here is an example of twine tensioning for a McCormick Deering baler. The spring tension is adjusted by loosening or tightening the nut.

EQUIPMENT OVERVIEW

Harvesting hay requires quite a few pieces of equipment. Besides the power unit mentioned earlier in the book, you will need something to cut the hay, something to ted or fluff the hay, something to rake the hay, something to bale it, something to transport it, and possibly some automated handling equipment. I'll go into to each briefly and an in-depth discussion will follow.

The piece of equipment that cuts the hay is most properly called a mower, but depending on the type of mower, you may see it referred to a mower, haybine, discbine, mower-swather, or sickle. Haybine is a trademarked brand name of a mower-conditioner, which is a device that mows and conditions (squeezes and crimps) the hay in one pass. A discbine is a mower with several rotary discs with small knives at the end. This type does not condition the hay during the mowing pass. A sickle mower is the style of mower that cuts in a scissor action and is the type of mower folks usually think of when they think of hay mower. A mower-swather is a mower that cuts and bunches the hay into a windrow in a single pass. This is actually more commonly used for small grain crops that are harvested green and chopped into silage. The most widely used mowers are mower-conditioners with sickle-type cutters and a rubber roller-type crimping mechanism.

Conditioners are devices designed to crimp and press the hay between two-spring loaded heavy rubber or rubber/steel rollers. The idea is to bruise the hay so that it dries faster and to prevent the coarse, heavier stems from holding water. While you can buy conditioners as a separate implement, most growers who decide they need to condition their hay will purchase a mower-conditioner.

This photograph shows the adjustment for the tine of a combination rake/tedder. This tine is in the raking position and the other hole is for tedding.

Tedders are special rakes that fluff the hay while it is lying in the field. The sole purpose is to facilitate the movement of air among the hay crop and to expose uncured hay. Tedding hay significantly improves the curing rate and its consistency throughout the hay. These rakes are typically basket-like rotating devices with long tines that mix up the hay.

Rakes are designed to do just that—rake up the hay. Side-delivery rakes are the most common, but vertical wheel (often referred to as pinwheel) rakes and horizontal basket rakes are common, too. Generally speaking, side-delivery rakes are the most expensive, but they clean the field the best and are usually the most widely available on the used market. Vertical pinwheels have gained popularity in recent years and are best used on flat fields.

Balers produce bales in two shapes: round and square. Round balers create large bales that look like rolls of toilet paper, except they are made of hay and stand 4 to 6 feet high. These are used primarily on cattle farms. Square balers create the traditional 18x14x36-inch square bale. These are used by horse and small livestock farms. Recently, a new size of square baler has made a large impact on the market. This is

As you can tell here, the color change of this orchard grass hay from green to a silver-green indicates drying is moving right along. You should ted this as soon as possible to increase air circulation, drying speed, and the consistency of the dry down.

This handful shows why drying can take some time. Coarse stems, clover heads, and weeds dry slowly, whereas the hay itself may dry quickly. Tedding twice might be required in order to speed the dry down of the thicker sections of hay and the coarse and trash species, such as clover and weeds.

the large square baler that creates bales up to about 2½ x 4 x 8 feet. They are not for small-acreage operations, but they may make an impact on some of our markets in about 5 to 10 years. There are several balers of odd design that you may run across, and there are also many options that are available for your baler. An example of unusual design is the Allis Chalmers roto-baler. It makes a small round bale slightly larger than a typical square bale and it is shaped like a Tootsie roll. In fact it is often called a Tootsie Roll baler. Examples of options are a silage chopper (chops green crops and bales in one pass) or a fine pickup head, which picks up finer bladed crops more completely, such as full stands of bluegrass.

Transportation and handling equipment that you might expect to see or need include a hay wagon or trailer to transport the bales and an elevator to help move bales up into a loft. This may include conveyors to move the bales from the wagon to the barn if the wagon can't be parked near the barn or the loft is large.

To judge dryness, snap a stem of grass or clover. While it won't feel or sound as definitive as a twig breaking, it should certainly bend and then fail in a manner that is consistent with being dry. To get a feel for it and to develop a sense of what is dry and what isn't, start testing stems as soon as you cut when you know it's green and keep testing as your hay dries.

A great technique to get a feel for dryness is crushing and rubbing the hay in one hand. You'll be able to feel the hay very well and see how it behaves in your hand. This is a very accurate test once you have experience. You should oven-test your hay to test for moisture. Once you learn how that feels and reacts in your hand, you can usually tell by instinct when it is time to bale.

This device on the tractor's PTO shaft is an over-running clutch. Any tractor without a live or continuous-running PTO must have one of these. This prevents the inertial energy of the implement from driving the rear wheels of the tractor forward.

PREPARATION

Preparation for harvesting means making sure your equipment is tuned and adjusted, all the maintenance is complete, and all arrangements and efforts have been made to line up storage and/or buyers. Also, you should be arranging for bale handling at the barn if you haven't already and make any necessary arrangements to hire extra field labor. Now is the time to estimate the number of bales you will get so you can estimate the amount of help. Your agricultural extension agent will be of more assistance here than I

can be, but 20-80 bales per acre is the range, with 20-45 being the most common.

If you haven't done so already, find a place to rent or borrow to store your hay if you don't have enough room on your farm. If you can't justify the expense of a building just to store hay and there isn't enough storage on your property, you will have someplace to store the bales that don't sell out of the field. Next, do as much marketing and presales as you can. When you have customers lined up, make sure you create a list of customers who want to save money and buy the bales out of your field and which ones prefer delivery. The ones who prefer to pick it up out of the field will need to be called whenever you start cutting so they have a day or two before they will need to show up at the field.

Most important, you need to train yourself on your equipment if you haven't done so already. First, make sure you have manuals. Manuals for almost all implements are available at dealers and on the Internet; eBay is a great source for original manuals. Check the appendices for sources. The maintenance requirements for the implement are included in the manuals. Follow all of the maintenance instructions and assume no maintenance has been done if you bought used equipment. Clean the equipment and then test run it by cutting, tedding, raking, and baling a small section of your field. This test run will help you get your machine adjustments correct. You should cure this test cutting and let it dry like regular hay so your baler settings will be realistic for your actual cutting.

Last, but certainly not least, make sure you familiarize yourself with all safety precautions found in the manuals and remember the cardinal rule: Never get off the tractor while the PTO shaft is spinning!

LINING UP BUYERS

If you are growing hay for the market, you should pre-sell as much hay as possible. Long before you make your first cutting, you should do everything possible to prepare flyers, talk to agricultural extension agents, visit local stables, and

There are contrivances on mowers to make a swatch that is narrower than the cut so the edges of the cut are kept clear of hay. Above you can see a well-formed, clean edge beside the cut, and on the left is an arrow pointing out the left hand sheet metal piece that helps to form this swath. On the right side is an identical piece, which does the important job of keeping mown hay away from the unmown hay. Your mower will clog at the outside of sickle sections if the edges along the cut are not kept clear by the swath boards or "swath makers."

Here are the two primary ways mower-conditioners feed hay into the mower. The first is a metal roller bar that puts tension on the hay by pushing it forward. This tension helps the hay stand still and straight for the mower. It works well for clover and other stemmy hay varieties. The second is the reel, and in this picture (below) you can see the teeth of the reel pushing the grass up into the rollers after it has pushed it into the sickle mower.

talk with local farm and equine suppliers in your area about your hay. While having to store unsold hay is a reality all hay growers have to deal with, preselling as much hay as possible straight out of the field should be your goal. This will minimize your expense and time in handling the hay and keep your storage requirements to a minimum. After a few years, you will have developed relationships with enough local hay consumers that eventually your operation will sell most all of the hay it produces straight off of the field.

Initially, you will have to do a lot of advertising and word-of-mouth marketing to get these relationships started. The three most common methods for generating sales are flyers, classified advertising, and cold calling. Flyers are probably the most effective and most likely to generate local sales. Businesses that cater to hay buyers are feed and seed stores, farm supply stores, equestrian supply/tack stores, large animal veterinarians, and hardware stores. Don't forget some other sources, such as the local agricultural extension office, farm cooperatives, and farm credit unions and other financial cooperatives.

Classified advertising is the second most helpful way to generate sales and will help develop sales leads to horse/livestock operations outside of your immediate area. Your local

Left: The route you mow determines the route of the tedders, rakes, and balers. These implements and the mowers themselves work more efficiently and safely on fairly straight rows. However, smooth even curves in a hay field are perfectly normal and standard practice. Just don't allow these curves to become too extreme.

Right: Here is an example of a perfectly mown swatch of grass. The edges are clear, the hay was thrown into the swath in a consistent fashion, and the curves are even and gentle.

newspaper is an obvious choice to place your classified ad, but there are many others. For example, the financial cooperative that has my farm mortgage has a classified page on its website. The North Carolina Department of Agriculture has a classifieds page designed specifically for hay buyers and sellers on its website as well. Other places for advertising are farming newspapers that may exist in your area and radio programs for farmers.

The last method, cold calling, is the most likely to be dreaded by haymakers. The reason is simple: You are going

to hear rejection. This method has the lowest success rate in terms of sales. However, I wholeheartedly encourage everyone to use this method because it will help you understand your potential customers as well as your competition. There is no substitute for visiting likely buyers, seeing their operations, and hearing their hay-buying experiences. You will get to know your market and potential customer base better and they will get to know you better. Also, horse farmers who have talked with you will remember you. They will not remember a flyer or classified ad. Often a lot of other great

Becoming efficient at mowing means learning some tricks. Clearing jams without getting off the tractor seat saves a tremendous amount of time if you can perfect it. Here we have a standard jam where thick grass gets balled up before going into the conditioner. In the first picture you can actually see it forming to the left of the vertical white bar. Next, the ball of tangled hay is apparent. Catching a jam early is key to easy unjamming. To remove the jam, lower the mower-conditioner all the way to the ground and back the tractor up until the mower is in unmown grass, as you see in the next two pictures (opposite page). This clears the grass from the sickle sections although the jam is still in place. Back up 10–15 feet, then inch slowly forward until the conditioner picks up the ball and spits it out the back. You can see it behind the mower in the last picture.

information will come out in conversation, such as the type of hay they like, what types of bales they prefer, who they are buying from now, and more. This is strategic information that will help you become a better hay producer over time.

WHEN TO HARVEST

Your hay is planted, you've waited through the winter, spring is here, and your field is flushed with growth. Your equipment is ready and your marketing and sales efforts have yielded a few customers. When do you actually harvest? The first consideration is crop maturity. Next comes weather. Extremely hot and dry conditions should be avoided, as should cool, damp periods. Hay dampened by rain is ruined and usually can only be fed to goats or cattle and can't be fed to horses. Likewise, nutritive value and palatability is negatively affected by too much sun and over-drying. In most areas of the country Mother Nature gives us only certain windows of time to harvest hay and each cutting has its own concerns. In the Southeast, finding three or four straight days of low humidity and strong sun for the spring

cutting can be difficult. The summer cutting can be interrupted by unpredictable rains or weather that is too hot and a sun that bears down negatively on hay. The weather for a fall cutting is usually the most cooperative in terms of rain, but heavy dew, morning fog, a weakening sun, and lower air temperature increase drying time significantly.

As to maturity, each species of hay and forage crops has a growth cycle that dictates the best time to harvest. The nutritional value of hay peaks at a certain point in that growth cycle, which is when you cut, weather allowing. Cutting

before or after your hay has properly matured creates a product that has less than optimal nutritive value. To complicate matters even further, certain growing conditions, such as a drought, will create the need to mow the hay to stimulate a new growing cycle, but the volume of the mown product is so low you can't harvest—you have to just leave it on the field. If the early summer is dry and hot, the second cutting is usually mown down and not gathered for a harvest. The spring and fall cuttings are usually fairly dependable, but sometimes even the fall cuttings can be sparse.

Typically, you will mow counter-clockwise, but if there are woods or fences, the first trip around the field should be clockwise. This will mow closely along the edge and provide a lane for you to travel on as you start your counter-clockwise travel. Here, the tractor is traveling along the first cut that was made clockwise to make the second cut, which is the first counter-clockwise cut.

Deciding when to harvest usually means monitoring and juggling the time windows created by growth cycle, weather, hired help, and personal schedules, not to mention other factors such as equipment readiness, anticipated harvested volume, fertilization schedule, and market conditions. Therefore, the best time to harvest is when all of these time windows and factors fall together. In the best of times, these windows and factors will all open up into a three- or four-day period that will create an ideal harvest of hay. Most of the time you have to make compromises and there is no shame in it. If your crop is not quite mature but the weather is terrific and a long spell of poor weather is anticipated, harvest a bit early. If the volume is not there because of a drought but the crop is maturing, don't stress yourself, your equipment, and your sanity to harvest 10 bales per acre. Mow it and leave it as a green manure on top of your plant

USING A MICROWAVE TO TEST HAY DRYNESS

1. To test the dryness of your hay, grab a 1- to 2-pound sample of your hay and accurately weigh it on a scale, using a paper plate as the tray for the hay. Exact size of the sample isn't important but getting the weight accurate is. Don't worry about the weight of the plate because the plate will be weighed every time.
2. Place the hay in the microwave, and dry on high power for about 30 seconds. Weigh the hay again. Note the weight.
3. Repeat step 2 until you get two weight readings that are virtually identical. Stop using the microwave and get out the calculator.
4. Subtract the last weight from the first weight. Then divide that difference by the first weight and multiply by 100. This is the moisture percentage of your hay.

A couple of notes:

- Don't grab a sample of hay early in the morning. The dew will skew the results.
- You are shooting for 15-20 percent moisture. Anything less will create leaf shatter and dust problems; anything more will mean hay spoilage in the bale.
- Remember, you will lose another 1-3 percent as you bale a field. If you test your hay around noon, and get a 20 percent test result, you are fine and can start raking. Most of the hay will be 17-19 percent by the time it actually gets in the bale.

Every hay field will have a sharp corner or two that can't be avoided. Here is how you operate a mower around such a corner. It seems instinctual, but it takes some practice if these corners are not to be a source of jams and poorly mown hay. Start getting ready for the turn by having your hand on the hydraulic control so you can raise the mower. As you get close, time the raising to exactly match the end of the row, and raise the mower without shutting down the mower. As you start the turn, shut down the mower. You shouldn't turn any operating implement sharply while it is operating, and turning sharp is necessary to make these turns efficient. Start the mower again and get your hand on the hydraulic control once more. Drop the mower when you reach the start of the row. If you time it right, the mower won't clog on mown hay because you dropped too soon and there won't be any unmown hay where you dropped the mower too late.

Pinwheel rakes or wheel rakes typically have the capability to alter the wheel height and approach. Height is obvious, but approach refers to the angle at which each individual rake is relative to the line of travel of the implement.

roots and don't look back. Perfection is an unreasonable goal when most factors are out of your control. Just remember, your decision to harvest should err on the side of weather when you are unsure. Hay ruined by bad weather is pointless. It is much better to harvest over-ripe hay than to harvest no hay at all because it was ruined by rain.

A fair question at this point would be when is the optimum time during the growth cycle to harvest? The answer actually varies quite a bit between the different hay and forage species. In alfalfa and clover, the proper time to harvest is when about 25 percent of the field has flowered. In grass hays, the right time is usually just as seed heads start to appear in the field. Of course an occasional plant goes to seed extremely early, but wait until you see that leaf growth has appeared to maximize and some plants are showing seed heads on close inspection. When alfalfa and/or clover have been sown in with grass hays (an orchard grass clover mix for example), typically the legumes mature earlier, but the grasses should dictate the harvest schedule. Grass maturation may get ahead of the legumes in the spring if the fall cutting was missed and spring growing conditions are ideal for grass.

There is a method of feedback for your decision making in terms of harvest time. Have your hay tested by your local cooperative extension service office. For a minimal or no fee, it will analyze your hay. Depending on the state, you will receive tremendous information about the nutritive quality of your hay. While this information measures many different aspects of your hay operation, such as how well you harvest and dry your hay, it also measures how well you choose harvest times. Low protein readings combined with high fiber readings and a notation of seed heads on the report could mean that you are waiting too long to harvest. Your local agricultural extension agent can help you interpret your report. The agent typically will also lend the appropriate hay sampling tools since bales of hay are sampled by coring instead of being grab sampled.

PERCEIVED QUALITY

From the hay growers' perspective, quality is the same thing as nutritive value. Raw nutritive value is the only quality measurement we can use. Color, weight, aroma, and other subjective or quasi-objective standards are not useful. From your perspective or that of your customers, the word "quality" is a complex word. Quality from the animal owner's perspective takes into account palatability, value for the dollar,

The picture above shows an example of what improper rake adjustment will do. The windrows are inconsistent and the rake isn't picking up all available hay and bringing it into the windrow. On the right is an example of a better raking job. All these deficits have been addressed, and the windrows are reasonably true and straight.

nutritive value compared with the life circumstance of the owner's particular animal, and more. To make matters worse, these stockmen and owners often use arbitrary and capricious methods for judging quality.

The two most common methods used by buyers to judge the quality of your hay are color and weight, and incidentally they are the least reliable indicators of quality. The fact of the matter is the weight of a bale has very little to do the quality of hay, although it could indicate a better value economically when compared with equally dried bales that are lighter (more hay for the dollar). However, weight usually has more to do with moisture content and not the amount of the crop. For example, a hay bale made in the fall will usually weigh more than a summer-made hay bale that has an identical amount of hay in it because fall hay typically has a slightly higher moisture content. Additional

Windrows can have sharp turns if needed. Here the windrows take a turn to the right to avoid a ditch.

weight doesn't mean the buyer is getting more or better hay; it is a bogus measurement of quality.

Color is another criterion buyers will use that is a poor judge of quality. While color can be helpful to the buyer in eliminating sun-bleached or over-dried hay, it has very little value any further than this. I can show you two bales of hay from the same field with very different colors that have nearly identical nutritional analysis reports. Buyers can, and occasionally do, use other standards of quality that are a little more rational, such as looking over your nutritional reports, texture (stemmy means less protein), and weediness.

While weight and texture are easily addressed by choosing harvest time correctly and not baling light (the practice of using loose bale tension to create a higher number of bales), using a color standard is a little more difficult. Here are some issues to consider. Fescue sun-bleaches quickly and will brown quickly. Anything you can do, such as additional tedding, to speed drying and get the fescue into a bale will help. Orchard grass retains a silver-green

While twin basket or twin reel tedders are very common on small acreage operations, quad-basket tedders are also widely seen and used. However, transport and storage is a problem for such a wide implement. Here you can see the hydraulic ram (the long silver bar is part of it) that raises and lowers the outside basket on each side, but some baskets are moved manually.

The side-delivery rake you see here is the most commonly used style of hay rake in most parts of the country. This rake typically builds consistent windrows and cleans the field well. It is easily adjusted and can be set up in tandem to double its working width.

color, but clover often turns black quickly as it cures. Alfalfa has a famous green color but will often gain a yellowish hint in wet weather. If you have had a lot of wet weather, let an alfalfa field get a few extra days of sun to deepen the green before harvesting.

Baling hay that looks good, is nutritious, is palatable to the livestock, is weighted fairly for the buyer, and enhances your reputation as a grower should be your goal. Balancing all these notions that surround the ideal bale of hay can be a bit tricky, but the rest of this chapter should help you through it.

HARVESTING OPERATIONS
Hay Cutting and Conditioning
Choosing a method to cut hay isn't as simple as one might think. This is because contrary to most beginners' knowledge and intuition, you are usually (but not necessarily) performing two steps during cutting. The first step during cutting is obviously cutting the hay. The second step performed during cutting is to condition the hay. To condition hay is to crimp the hay to improve its drying speed. There are two different methods used for the first and primary step of cutting hay: sickle mowing and disc mowing. Sickle mowers are the implements that mow hay with the traditional scissor-style action. Disc mowing is a newer style of cutting hay that makes use of high-speed spinning discs that have two small steel flail knives 180 degrees from each other. The working width of your sickle mower is the same width as the sickle. For disc mowers, the working width is determined by how many discs your disc mower has. While disc mowing is

gaining in popularity, I believe it will be several more years before enough disc mowers are on the used implement market to make inroads into small-scale haymaking operations. New disc mowers and disc mower conditioners are clearly beyond most small-scale operations budgets.

Please note that using a brush mower or Bush Hog is never an acceptable way to cut hay. It cuts the crop in a ragged fashion, which is hard on the hay, chops much of the hay into smaller pieces, and leaves the hay clumped in a way that inhibits drying.

The conditioning step is usually done with cutting machines (either sickle or disc) that feeds the freshly cut hay into a combination of mated rollers (usually it is just a pair but I have seen three) to crimp and press the hay. There are

two different machines that will both mow and condition the hay. The first is a mower-conditioner, and the second is a disc-conditioner. They are both correctly called mower-conditioners, but in common use the phrase "mower-conditioner" is reserved to the machine that combines a sickle mower and rollers to condition the hay. This sickle-style mower-conditioner is more widely known as a Haybine, the New Holland trade name for its mower-conditioner. The second machine is a disc-conditioner and again is often referred to as a Discbine, the New Holland trade name. This machine combines a disc mower for cutting and roller technology for conditioning.

All types of mower-conditioners use rollers for conditioning the hay. Most often they use rubber rollers only, but often older mower-conditioners have one steel roller mated to a rubber roller. Both types can be used for hay, but if your hay is a coarse, warm-season grass, such as bermuda, mixed with other coarse forage like clover, you may find that a

These two pictures show the adjustments for combination rake-tedders. The first photo is of the tines themselves. The two positions for the tines determine tedding and raking, respectively. The gates have multiple positions depending on the volume of the hay in the windrow and how tight you need the windrow to be.

mower-conditioner with at least one steel roller will condition your crop better than only rubber. Growers with only fine grasses like bluegrass and fescue will find that only using rubber roller conditioners will adequately crimp the hay and be easier on the crop. When buying a mower-conditioner, make sure that the rollers are not severely damaged and are mated with sufficient tension.

There are other alternatives to these combination machines. The most commonly used by small-scale operations is a dual-pass mowing-conditioning. First there is a mowing pass and then a conditioning pass. The mowing pass is commonly performed by sickle mowing, but a disc mower can also be used. A separate conditioning machine then passes through the field to condition the hay. Since mowing/conditioning productivity is cut in half by having to make two passes, this isn't a popular method. However, this same lack of popularity means machines that just condition or mow are separately much cheaper than mower-conditioners or disc-conditioners. If your hay operation is smaller than five acres, you may want to consider dual-pass mowing and conditioning. If your acreage is much larger than this, you may want to reconsider two-pass mowing/conditioning because of the large amount time it would involve.

Tedding

Tedding is the act of fluffing the hay to increase the drying speed of the hay. It helps improve the consistency of the final moisture content of the hay. Tedding is required in most climates, and to understand why it is required, you have to understand the hay drying process. Hay dries through the direct escape of moisture from the plant material through the leaves and the ends of the stem. The moisture escapes through the action of direct sunshine (radiational drying) and air warmth (convective drying). Hay will give up its moisture immediately after it is cut. In the first 12-24 hours of drying, warm air, low humidity, and a gentle breeze is all that is required to create significant drying.

However, as hay dries, the rate of drying through convective means slows down and the only way to speed up drying is to add energy to the plant material in the form of sunshine. That is where tedding comes into play. It mixes up the plant material to make sure all material is evenly exposed to sunshine and improves air circulation, which will also help speed up the process of drying.

Now that you understand the drying process, you understand that the goal of tedding is to make sure that all your hay is evenly and completely dried. As important as tedding is, especially during the cooler cutting times of the year, you have to be careful not to over-ted the hay. Leaves of the most grasses and legumes lose strength as the plant dries, which makes it easy to knock off leaves and tear apart the hay as you ted. This is called leaf shattering. As much as 10–20 percent of your production can be lost by tedding too vigorously or too late into the drying process. Alfalfa is especially susceptible to damage from over-tedding or tedding too late in the drying process.

There are two primary implements that are widely used for tedding, and I have used them both and had success. A basket-style tedder is driven by the PTO shaft of the tractor and spins two or more horizontal wheels that have thin, spring-loaded teeth pointing down to the ground. The wheels spin fairly quickly and the hay is flung from the rear of the implement to scatter and mix the hay. These types of tedders do a great job of tedding but can be rough on the hay and create quite a bit of damage.

The second type of tedder is a rake-tedder. Rotary rakes, which look like a tedder with large baskets and adjustable teeth and gates, ted hay like a rotary tedder but

Here is an example of the raking technique you'll need to use if your harvest is sparse. In the first picture, a windrow is made from each swath. However, if you don't have a lot of hay, you'll need to combine two windrows to provide the sort of volume the baler needs to make well-formed bales and provide some measure of efficiency.

The crank on the top of the tedder in the first picture is how you adjust the angle the rotating baskets relative to the ground. This angle is called the implement's presentation. In the second picture, the presentation is adjusted through the top link, as with most three-point hitch implements. The second picture also shows the tedder-rake setup with the proper presentation.

can also rake hay. Likewise, there is a side-delivery version of the rake-tedder that is designed to operate as a tedder. It does this through the orientation of the teeth and by running backward. While running as a tedder in this fashion, a side-delivery rake will gently roll the hay over instead of raking it up. Standard basket-type tedders are widely used and work the best in my estimation. Combination rakes work well and have a financial benefit. Side-delivery rake-tedders are gentler to the hay but they may not fully ted the hay.

Raking

The act of raking hay is really an act of baling preparation or pre-baling. All you are doing is getting the hay ready for the baling machine. Raking, unlike tedding, which has a direct impact on the quality of the hay, provides no benefit or improvement to the hay itself. Some growers will allow the hay to continue to dry after the hay has been raked into long lines (windrows) in the fields. This isn't generally accepted or even proper in many climates. The act of drying hay while it is in the windrow is called sweat-

This picture shows why you ted. The swath to the left indicates the compaction of the crop by the travel of the mowing equipment and the mower-conditioner itself. The swath immediately behind the tedder shows how the crop was lifted and fluffed, and allowed more complete and consistent drying.

Tedding should never injure or flail the crop. This picture shows proper technique. The hay is being turned and scattered but it isn't being damaged through excessive flailing This is achieved with the right combination of PTO speed and ground travel.

ing the hay and should be avoided unless directed by your agricultural extension agent (very dry climates can avoid sun-bleaching and over-drying by sweating). Climates or cutting seasons with heavy morning dews may find that putting the hay in a windrow the night before baling may speed up burning off the dew and give you a few more hours of baling time the next day if the hay is sweated in the windrow. Generally speaking, you shouldn't rake until the hay is ready for baling and it should never be put in a windrow until it is dry.

Raking is fairly simple, and there isn't much you can do to affect the windrow size and density, which are dictat-ed by your rake and hay. However, you can dictate how thoroughly the field is raked. This is accomplished prima-rily through vertical height adjustment of the rake. Height adjustment is accomplished through a leveling crank on one or both land wheels. There is no proper height in terms of a particular measurement. You should adjust the rake based on performance. The rake is properly adjusted if the rake teeth operate close enough to the ground so that vir-tually all hay is brought into a windrow while maintaining enough clearance between the teeth and the ground to avoid regular and repeated contact with the ground and harming the crown of the plants and/or the rake.

While raking, there are a couple of other ideas to take into account. Balers create uniformly dense bales if the windrow is full and consistent. While creating perfectly full and dense windrows is obviously impossible, you should work towards that end. You can do this by combining two or more thin, weak windrows into one windrow; avoiding sharp turns that tend to scatter the windrow; and keeping your forward travel speed consistent and not too fast. The more experience you have with your baler, the better you will become at raking.

Baling

If you have gotten to the point where you are ready to bale hay, you have been quite successful so far. A lot of pitfalls and weather-related obstacles could have prevented you from even getting here. It is not unusual for heavy rains to wash out new plantings or ruin hay curing in the field, equipment breakdowns that leave hay in the field too long, and more. However, like all harvests, until it is in the barn the work and worry isn't over. It is time to get it off the ground and into the barn.

Traversing small ditches is sometimes necessary when doing field work. Unfortunately most haying equipment crosses ditches very poorly. While many mower-conditioners can be raised high enough to do it, balers typically can't cross ditches at all, and rakes and tedders are problematic. Here we see the proper technique. Never cross a ditch with the PTO engaged. Approach the ditch straight on, and never at an angle. The tractor crosses the ditch slowly and carefully, one axle at a time, and then the rake or tedder is eased over the ditch.

This is an example of how to adjust the gates of a rotary rake/tedder when tedding. They are set wide open to allow full scattering and turning of the hay.

When turning a three-point hitch implement, realize that turning radii are often diminished by the hitch frame or geometry. In addition, the offset nature of the implements creates a situation where turning radii differ depending on which way you turn. This three-point tedder/rake turns well to the right, but it has double the turning radius if turned to the left. You have to plan your raking so you can travel clockwise to take advantage of the rake's differential turning abilities.

A side-delivery rake operates by gathering the hay and moving it to one side; hence the name. This movement tends to roll the hay into a well-formed windrow that feeds into a baler without jamming. In fact, in terms of windrow formation, a side-delivery rake is tough to beat.

While it is easy to say that getting uniformly dense and properly weighted bales is the goal, achieving that goal is pretty tough. Doing this well reveals the art in the operation of hay equipment and hay farming. An experienced hay farmer can operate the equipment in such a way that the rate at which the hay is fed into the baler is perfectly matched to the baler's capacity and plunger stroke speed. This delicate balance creates a uniform and properly weighted bale. It also requires a lot of practice and experience with the equipment.

How do you balance all the factors to create a perfect bale? There is no magic formula. You will have to work out

a system on your own; however, I can give you one rule of thumb. I have found that I achieve consistency by making sure each bale has the same number of baler plunger strokes.

If I manage this one parameter above all others, I am more successful at creating uniform bales. This is not to say you shouldn't worry about ground speed, windrow size, and other parameters. Counting plunger strokes and getting a feel for that will help make the other parameters more manageable. If the hay has been consistently dried and the windrow size is fairly consistent across the field, then using plunger stroke count as the guiding parameter will create amazingly consistent bales. In other words, for any given

Here is an example of how to load a baler. The windrow isn't too large, but it isn't too small, and the baler is also getting plenty of hay to work with. Most of the hay is getting fed to the outside of the pickup head, which is okay, though a more consistent feed across the entire pickup would be ideal.

Setting the pickup head of the baler is very important. If it is set too low you'll break drive chains or the drive belts will slip as the times intersect the ground hard. Too high, and you'll miss the shorter and finer hay and reduce yield. Experience will quickly show you what to use in your circumstance, but always err on going too high because equipment breakage isn't worth getting every last piece of hay in your field.

bale tension setting, counting plunger strokes is more effective for consistency.

I'll use my baler setup as an example. An average bale weighs 50-70 pounds and has about 15-20 flakes (slices, pieces) in each bale. Since one plunger stroke represents a flake, I shoot for 20 plunger strokes per bale. I then set up bale tension so I can easily create 3-pound flakes using typical ground speeds. This will create something on the order of a 60-pound bale. I write this as if it is easy and simple, but nothing is farther from the truth. Managing these factors is the art of haymaking, and practice makes perfect.

These two cranks raise and lower the long middle bar and are for setting bale tightness. The more you turn the cranks clockwise, the tighter and heavier the bales will be. If the cranks are too tight, you'll break twine and create bales that are too heavy and could possibly be a spontaneous combustion hazard if your hay is wet. You are shooting for 40 pounds of dry material in the bale, so if your hay is 17 percent moisture, set your bale tightness to give bales that weigh just shy of 50 pounds.

Most balers only create bales that are as consistent as the feed rate and style with which the baler was fed. Here are four examples of improper loading. First, all four windrows are too light and don't have enough hay in them. The third is being fed inconsistently and the other photographs show biased loading (top, middle, and bottom respectively). Bias and inconsistent loading and light windrows lead to short and/or lopsided bales in most balers.

There are a couple of other tricks, too. While the baler is designed to run at 540 rpm, it won't hurt it to run at 480 or 600 rpm, and adjusting the PTO speed slightly using the throttle can be used to overcome a limitation you may have with forward travel speed. A tractor or continuously running PTO is a requirement for effective hay baling. If you have live PTO, you can let the drive clutch of the tractor in and out as needed to maintain a very slow creep through a particularly thick, heavy section of windrow. There is a very slight, gentle grade in one of my fields, which also happens to be one of my most productive sections of hay and the windrows are almost always thick. I simply push the clutch in and let gravity inch the tractor along the thick, heavy windrows. Just remember that safety is more important than anything else, and ground speed must first safely match terrain and ground conditions.

BUYING HARVESTING EQUIPMENT

Buying used equipment is like buying produce. What you have available to purchase may not be ideal, but if you are

Here is an example of the two knots tied by the vast majority of balers. The first is a loop knot, which is an overhand knot tied around the base of a loop. IH, Allis, and older grain binders tie these types of knots. The second style is a modified square knot, and the most common baler that ties these are New Holland balers.

Most balers have a simple drop chute at the end for discharging a bale. Others, such as this New Holland, have additional options for dropping bales to the side, doubling bales before dropping them to decrease the number of bale pickup points by half, or have a bale-kicker attachment that launches the bale up into the air and backward onto a wagon being pulled behind the baler.

Since balers are offset, traveling down the road with them would be tough if the tongue weren't adjustable. This rope pulls the pin and allows the operator to adjust the tongue to the side. This rope should be removed before operation to prevent tangling in the PTO shaft.

going to eat vegetables you better buy something. This is a small, barely sufficient analogy to illustrate that small-scale hay growers suffer the vagrancies of the used equipment market and these quirks of fortune dictate what is and isn't available to us. Therefore, most of us acquire rather eclectic collections of equipment, none of the pieces sharing a common manufacturer, time of manufacture, or design similarities. We simply put together the best set of equipment we can and continue to trade and buy until we have a set of equipment we are comfortable and happy with.

As your collection of used equipment evolves and grows, you should strive to put together a system of implements that work together in similar working widths and require similar tractor sizes to operate. While exercising brand loyalty by putting together a set of equipment from the same brand isn't something that makes sense for most of us, it does make sense for some, especially if a particular brand dominates the market in your area. Another reason may be nearby dealers. If there is a dealer for the brand you are considering to purchase that is still in business in your area, this dealer represents a significant body of knowledge,

When deciding on equipment, you must choose between draft-style implements and three-point hitch implements. A Ford tedder with a three-point hitch with hydraulic folding extensions is in the first photo. In the second, a draft-style tedder with manually folding extensions is featured. The draft height adjustment for the three-point hitch determines operating height of the implement in the first photo, while a crank adjusts height in the second picture. I like to stay away from three-point hitch implements because draft implements tend to be much easier to store, adjust, hitch to the tractor, and they tend to stay in adjustment in the field much better. They are also typically cheaper. However, I know many who disagree, so the choice is yours.

support, and parts that may one day mean the difference between getting a part and getting back in the field to get your hay up before a rain or waiting for a part that leaves some or all of your hay unharvested and wet. For example, there are several Case/IH dealers in my area that can provide me with most of my baler's parts even though the baler is more than 40 years old.

The next consideration is the size of equipment you will be using. Tractors that are affordable and widely available to small-scale haymaking operations can usually muster the horsepower necessary for operating equipment that operates at or near 10 feet wide. Typically speaking, a 30- to 45-horsepower tractor will capably operate a mower-conditioner with a 7- to 9-foot cutting width, a 10-foot rake, a 10-foot tedder, and a baler that can comfortably bale from a windrow created by a 10-foot rake. If you live in an arid region with sparse hay fields, you can operate wider equipment. Irrigated hay fields and thick fields of hay in bottom land may dictate slightly smaller implements. Types of equipment will alter this slightly. For example, different types of rakes vary significantly in their horsepower requirements. Pinwheel rakes require very little power to operate,

A hay elevator is required if you are storing hay in a loft. You can usually pick these up at farm auctions for a decent price.

A wheel rake is a popular choice for many folks. This is a large hydraulic double boom. These rakes are simple and require no power because travel along the ground forces the rake wheels to turn. These are the cheapest rakes per foot of working width. The second wheel rake is more in line with a small haymaking operation. It is three-point-hitch mounted, but is in scale with our budgets and typical field sizes.

while high-speed tedder-rakes will require more horsepower. Again, local advice is very helpful here and the used market will probably dictate exactly what size equipment you should buy.

Once you have made your brand and size decision, you should decide what types of haymaking equipment you want. Certain types of equipment, such as balers, differ very little in significant ways between brands and models. However, there is quite a bit of difference in styles and type of mowers, conditioners, rakes, and tedders. The type you buy depends slightly on the brand you decide to buy, but most brands did make many different configurations of implements.

There are two primary types of tedders: tedders and tedder-rakes. Tedders are usually shaft-driven and are available in different working widths based on the number of

Left: When buying equipment, make sure the equipment is complete. This grain drill's gates are hydraulically operated, but the hydraulic ram that is supposed to exist between these two points is missing. The price looked like a deal until you factored in the cost of installing your own hydraulics. Right: This John Deere 24T baler's mechanisms are a combination of shaft, chain, and belt drives. The main timing is shaft driven from the transmission. These shafts (to the left and bottom of the picture) drive the chain-driven knotters. The second picture is of the belt-driven feed auger. Some balers are purely shaft-driven and others are mostly chain-driven. Stay away from balers that are primarily chain driven, but a machine that is primarily shaft-driven, like this one, is still a good choice. In fact, the 24T is a respected baler that works well on small acreages.

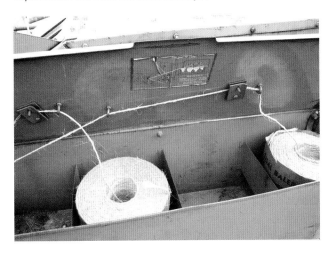

Twine routing and tensioning setup varies quite a bit between baler models. The John Deere baler uses two tensioning plates and eye bolts, while the New Holland uses a shared tension plate and separate brackets with grommets for each possible twine spool setup. The IH uses one tensioning plate and shared guides for both twine leads. The latter is the most straightforward and easiest to set up and adjust.

horizontal baskets that are mounted on the main frame. Small-scale operations should look at tedders with working widths up to 15 feet and should look at possibly buying a piece of equipment that serves as both a tedder and a rake. This will decrease your capital expenditures significantly and most tedder-rake combinations work reasonably well.

Combination tedder-rakes look like and operate at much of the same speed as tedders. With the teeth in tedding condition and the gates (devices that determine where the

Here are the two most common styles of conditioning or crimping rollers. The first (top) is segmented rubber over steel, which is cheaper to repair and may be harder on some types of forage and hay. The second (middle) is two hard rubber rollers, which is more expensive to replace and repair and may not fully condition coarse hay.

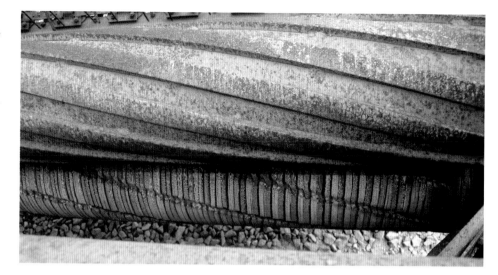

When buying a conditioner, make sure the swathing sheet metal is present and in good shape. It is quite common for this sheet metal to be bent or severely rusted.

Here are three examples of how springs offset the weight of a mower-conditioner so the conditioner can float along the ground. Ideally the weight of the machine in the cutting position has an effective weight that is only slightly positive. The machine will lift itself with every bounce or bump, but settle back down quickly afterward. Both designs work well, but the New Holland and John Deere are easier to adjust than the IH, and their springs are more out of the way for maintenance and repairs. All three adjust the tension using nuts at the end of threaded rods through the center of the springs.

hay is thrown) completely open, they will ted hay. However, with the tines adjusted in the raking positions and the gates closed or narrowed, the hay is raked into a windrow. There are side-delivery rakes that will also act as tedders. These are wonderful, but are quite expensive and can be difficult to find. For small-scale operations, the trusted basket style tedder and a side-delivery rake is probably the best combination. However, check with a couple of haymakers and implement yards and see what folks in your area like to use. Vertical pinwheel rakes are gaining in popularity in many areas.

Balers are very similar, and there are not many differences among them. You will probably hear all sorts of old

Here are three different power delivery systems used on side-delivery rakes. The first is PTO, which are usually belt-driven, as this one is. The second is ground-driven (the wheel drives the transmission, which drives the rake) and is the most common. The third is driven via a hydraulic motor. This is uncommon, but it works well for very fine control of rake tine speed.

A complete new set of tines for a rake is actually quite a bit of money. This used rake that is for sale has new tines, which makes it a better buy than one that hasn't had the tines replaced.

wives' tales about which brand or model reliably bales hay. These will include all sorts of stories about which model doesn't miss many knots or have many twine breaks, won't slug the hay, and feeds hay continuously and evenly, and more. The bottom line is that most older balers that are still around today and are field ready are proven balers by definition. Any of these balers, if properly maintained and adjusted, will bale hundreds of bales a day with only a few malformed, untied, or broken bales if properly set up. As you look at balers, have the seller operate the bales for you before you buy one. Spread some hay or straw out on the ground in a row and bring your own bales of hay with you if you must. Be sure to have the seller operate the baler to your satisfaction before you purchase it. If he or she refuses, pass on the baler or buy it for nearly nothing because you have to assume it won't reliably bale if the seller won't demonstrate its abilities.

˝   When buying a baler, there are several points to consider. One is general care. A baler can't be stored outside. Temporary storage under a tarp is okay, but don't buy any baler that doesn't look like it spent most of its life under a roof. Look for evidence of fastidious maintenance. Look for oil on chains, grease fittings that have evidence of greasing, and chain and/or belts that have been properly adjusted. There are few common problems to be on the lookout for, such as broken shear pins.

As to baler design, there are really only two designs to consider: chain-driven and shaft-driven. Shaft-driven balers never get out of sync with themselves unless something

This chisel plow can be used for subsoiling your pasture. This creates deep slices that improve moisture retention and soil aeration. Chisel perpendicular to the fall line of your land to maximize the benefits. While it is typically done in addition to plugging, it is completely optional. Some soils benefit from it and your agricultural extension agent can help. This is one implement where a three-point version is preferable to a draft-style subsoiler.

breaks. Every component works in concert with the next because they are attached to each other via a shaft so they can not vary in terms of timing. However, chain driven components can get out of sync. For example, the chain could skip a tooth or the tension of the chain changes over time. For this reason I recommend shaft-driven balers, which fortunately are the majority of balers available on the used market.

The pickup head is an area of concern when dealing with balers. This is the part of the baler that pulls hay in

Side-delivery hay rakes are suspended devices—the rake itself has a complete suspension system that is independent of the frame. These three arrows show the spring-mounted rods that hold up the rake assembly and allow the rake to travel up and down as needed. While the idea is to adjust the rake so no part of it, other than the tines, constantly touches the ground, the reality of farmland is there will be a few bumps and ditches that require the rake to give as the frame travels over the rough spots. The second shot shows a close-up of one of the spring-mounted rods. Make sure they are not rusted and move freely.

from the windrow. The head floats up and down and follows the contour of the ground. Because of this, there is often some slight damage where the head hit a ditch or rough ground broke the pickup tines. The head should have a few stops that limit travel and it must not be broken. No more than a few tines should be broken and the head should show no signs of welding or other major repair.

Another area to pay close attention is shear pins. There are several shear pins on most balers, but to illustrate what type of problem to be on the lookout for, I will use the flywheel shear pin as an example. The job of the flywheel is to smooth out power delivery from the tractor, and the flywheel is actually quite large on the baler as far as flywheels go. Because of the stored energy in these flywheels, shear pins are designed to break if the baler jams. Unfortunately, a lifetime of sheared pins creates a shear pin holder with quite a bit of wear. This wear may be extensive enough so

that the shear pins are always loose and frequently shear. The proper fix is to weld and reshape the hole or holder, but often a frustrated or careless farmer will weld the flywheel to the shaft flange and eliminate the shear pin all together. This repair will ruin the baler the next time the baler jams. Be sure to look at for shear pin elimination welds. There are usually shear pins for the bale feeder and they are sometimes used for the auger feed.

The bearings the plunger arm rotates on should also be inspected. Make sure there isn't any excessive play. Many balers also have back-feed plates in the bale chute that prevent the bales from sliding back and forth on the bale chamber as the baler builds the bale. They are sort of like barbs or undulations on plates that are mounted in the bale chamber. These barbs wear and the plate will need to be replaced. Tires, chains, and twine grommets (doughnut-looking object that the twine passes through) that are present and smooth, and

The New Holland rakes typically have hand cranks for adjusting the height of the side-delivery hay rakes.

This is a close-up of the New Holland suspension system. This system allows for side to side, forward to backward, and up and down movement. It is made of a slide track on the front and a bracket on two rotating cuffs. The springs that cushion up and down movement are the leveling crank springs.

a pickup head that doesn't show abuse or an excessive number of broken teeth round out the baler inspection.

As you shop for implements, there are two simple rules. Balers, mower-conditioners, and rakes must never have been stored outdoors for any length of time to be a viable working piece of equipment. Even tedders should not show signs of excessive outdoor storage. The second rule that is nice, lightly used hay equipment is hard to find. You should begin shopping as soon as possible. If you are not experi-

enced at looking at used equipment and are not sure exactly what a solid, lightly-used piece of equipment that has been well-cared-for looks like, look at a lot of equipment before you decide. After you look at a dozen or so pieces, you will be able to spot the pieces that haven't been abused, left in the rain, or poorly maintained. There are nice pieces available that will give years of service with only standard maintenance and typical repairs if you are patient.

Auctions are your best source to find field-ready and

complete collections of equipment, but very nice one-off pieces are best found in agricultural classified advertising. As you begin sowing seed, start shopping for harvesting equipment—start earlier if you plan to plant in the spring. Equipment is the most available during the season when you use it and are at their best prices in the off-season.

STORAGE

Hay can be safely stored anywhere that direct sun or rain can not reach it. In humid climates, a completely enclosed building is preferable, but in more arid climates, any roofed structure, with or without sides, is fine as long as direct sun won't reach it.

Hay will continue, albeit very slowly, to cure in the barn. Hay tightly stacked in a barn without air spaces between the bales or stacks will lose quality over time through excessive heating and mold and fungus growth. This is true regardless of how perfectly the hay was dried in the field. To delay the loss of quality that occurs during storage you must allow for some air circulation, which can be done by properly stacking the bales.

To allow some circulation, stack the bales so that there

Many balers use track-mounted feeding forks to feed the hay into the bale chamber or the fork is used in conjunction with a feeding auger. Either way, the tracks need constant inspection because they are prone to damage. This baler has a bent track. While it can be easily repaired, it is something you will need to be aware of before purchase.

Left: When bales are ruined, you can chop them up for compost, cover, and more. This bale chopper makes short work of turning hay into soil cover or material fine enough for the compost pile. Right: How do you decide between two implements that are similar in the important respects? Look for the touches of engineering that make your work safer or easier. The round tube seen here between the frame of the tedder and the extension has a spring in it, which makes lifting the extension very easy and might save the cost of having to go hydraulic.

If you make round bales, a method for transporting the bales is important. This three-point hitch-mounted baler spear works great for that purpose if you have a large enough tractor. Remember that smaller tractors may not be able to carry them safely, even if they are able to lift them.

is a small space between every bale. This space doesn't have to be large, just a finger or two wide. Unfortunately leaving spaces between the bales can make for a weak stack of bales. The stack will be more stable if the bale lengths are the same length and a little bit longer than the normal 36-inch-long bale. While you can't make perfect stacks of hay, creating bale lengths that allow for air space should be your goal to help ensure that the quality of your product last as long as possible while stored.

Loss of quality during storage is significant. You should try to use or sell the hay within 6 months of storage. After a year, the nutritive value of the hay has diminished to the point that the hay probably shouldn't be sold or used except as emergency or charity (e.g., donate to equine rescue organization). One important note is that extremely wet bales will mold and decompose quickly enough to generate

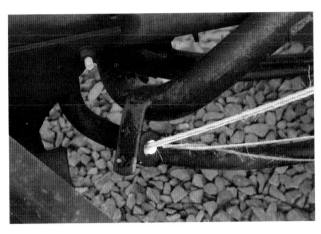 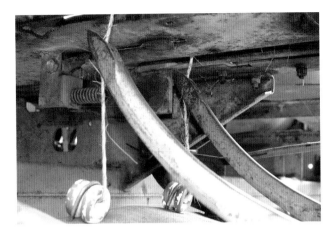

Left: When inspecting balers, look for damage to the twine routing guides and their brackets. Specifically, look at the guides found on the needle arm and the two guides under the baler. These tend to be damaged and are often overlooked by owners. Right: This baler's needles are in good shape, but look for signs that indicate if needles were broken and welded back together. While needles typically break, they can bend even though they are cast pieces of metal. Look for signs that the needle was heated (heat discolorization on part of the needle is the best clue). This will indicate the needles were bent and then straightened.

When inspecting the knotting mechanisms of balers, look for signs of excessive wear and evidence of neglect. Unfortunately, the New Holland baler to the right shows both, whereas the John Deere to the left does not. The first arrow in the New Holland photo shows wear on the shoulder of the crow's-foot shaft gear, and the other arrow shows lack of maintenance because the grease fitting shows no sign of grease. Compare that with the John Deere that shows neither of these problems.

One of the most reliable indications of whether a baler you are looking to purchase has had serious damage is a damaged bale knife. This one is in great shape, but look for heavy gouging, scorings, or fractures. I have even seen a baler for sale that had a shattered bale knife. Also, inquire as to why it was replaced if you find that the knife is brand new.

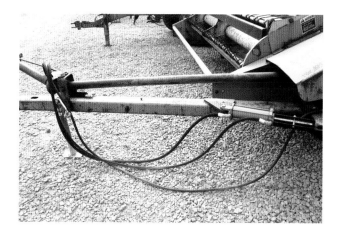

Transport and storage of large implements requires that the implement fold or alter its presentation at the very least. Most balers and mower-conditioners have a swinging tongue. In these two pictures are examples of the two most common designs. The first is a hydraulic tongue on a John Deere mower-conditioner. The hydraulic hose for the tongue's cylinder can be tucked away during mowing. You don't really need three remote hydraulic ports to use the implement. The second is a manual pin system. Pulling the pin with the attached rope and backing up or pulling forward as needed swings the tongue. The lever and roller design of the pin puller is a testament to the effort usually needed to dislodge the pin.

When deciding on balers, look carefully at the pickup head system. Here are two balers, one with a gauge wheel that allows the head to follow the land and the other has a rod and stop to allow the head to raise up if the head hits an obstruction. The gauge wheel design typically is easier on the pickup head and more reliable.

significant heat. This in turn will create a spontaneous combustion fire hazard (as the bale heats up the heat will set adjacent drier hay on fire). Leaving air spaces will allow heat to escape and prevent heat build-up should you have wet bales to stack.

HANDLING

Moving a couple thousand bales of hay a year can be tough, but it is manageable until you have to move the hay up into a barn loft. At that point, throwing the bales or carrying them up a ladder is typically impractical, unsafe, or both. At this point, some sort of hay conveyor or elevator system has to be used. The most common form of hay-conveying equipment is the hay elevator. The elevator frame is placed between the loft opening and the ground

Left: Fertilizer distributors come in many shapes and sizes. This is a handy size and configuration for small acreage. The draft-style hitch and ground drive makes this great for smaller older tractors without three-point hitch systems. It also operates the same as most lawn-type broadcast spreaders. The three-point hitch version is widely used as well. Right: One thing to look at when comparing rakes is if you have to remove the cross shaft to remove the tines. Many makes and models require you to thread replacement tines onto the cross shaft and force you to remove the cross shaft. One bolt for on-and-off replacement is a better design.

like a ramp. The elevator has a chain with small spikes that rotates around two axles—one at top and one at the bottom of the frame. The motor or PTO shaft drives one of the axles while the other axle is an idler axle. The hay is placed in the frame and the chain with spikes grabs the bale and moves it up the frame into the loft. The stronger the motor, the more bales that can be on the elevator at the same time.

If you are inclined to use a conveyor in the loft, commercial-style roller conveyors are your best bet. There are many styles of conveyors used in the loft and typically are roller-type conveyors. There are motor-driven conveyors that operate on the same principle at the chain-driven hay elevator. Occasionally you will find belt-type conveyors. Industrial, material handling, and other commercial auctions and used-equipment sales frequently have conveyors, but there is usually a better selection of roller and belt conveyors available. Chain-style conveyors aren't used often commercially and are better purchased at agricultural equipment auctions and sales.

The smaller elevators and other conveyers are typically driven by an electric motor. The larger elevators, especially those that are part of an entire conveyer system are typically driven by the tractor's PTO. If electricity is unavailable, PTO versions of smaller elevators are available, though buying one with an electric motor and then running it off a 120-volt AC inverter that is powered by a pick-up truck battery is preferable to purchasing a PTO-powered elevator. This will prevent you from having to tie up a tractor for handling hay during haymaking.

Above: You've made a bunch of hay bales, now what? Hopefully you answered that question earlier and have customers picking it up out of the field or enough storage space on your farm. Below: With small-acreage operations, usually two hands and a driver is all that is needed to help get your hay up. This hay is being taken to a barn at a location that requires travel on a public road, and is being loaded onto a street-legal trailer with functioning lights. Also, this trailer is handy for moving equipment and delivering hay to customers. Since you will find a regular road trailer handy for so many other reasons, you may want to consider purchasing a road trailer that is large enough for all your needs instead of a traditional hay wagon.

Left: Here is a traditional hay wagon that can hold about 150-175 bales. The sides allow the hay to be stacked high enough to travel along the road without a concern for lost bales and it's reinforced with a chain along the top. Right above and below: If you have access to reasonably priced lumber, buying the bare frame of a hay wagon and building the wagon yourself is usually cost effective. Here are two examples of bare frames. The two-axle frame will be rated for about 100-150 bales of hay, depending on the weight of the wood and bales. The tri-axle frame can handle up to 200 bales or more. Typically, the weight rating of the trailers exceeds the tractor's or truck's ability to stop and handle them, so load them with the number of bales your tractor can handle, not the number of bales the trailer will handle. Be careful when loading the trailer. Make sure you can stop it with the tractor's or truck's braking system, especially on grass and dirt.

TIPS AND TRICKS

If you are planning to sell your hay, the business is relationship driven and often long-term commodity-supplier-style relationships with your customers are created as a result. You will probably eventually sell all of your produce to a few farms that have come to appreciate and depend on your product. This arrangement is ideal for the small producer and the marketing circumstance is something you should try to achieve. It is also more cost-efficient and less of a hassle for you if your customers pick up their hay straight from the field.

While tedding helps to dry hay, you can over-ted your hay. The extra tedding may knock leaves loose from the stems, especially during the last stages of drying. This is especially true for alfalfa.

Sometimes the smallest adjustments matter. For example, my baler will start mistying bales as the knives begin to dull. Spending a few minutes every spring sharpening the knives will save hours of rebaling malformed bales.

Your grease gun is your best friend. I have never seen a piece of haymaking equipment that didn't have a large number of grease fittings and didn't benefit from frequent greasing.

Using high-quality twine saves much aggravation from mistied bales. Use the best sisal twine you can find to avoid having to rebale mistied bales.

When joining two spools of twine, use a simple double knot and pliers to help make the knot tight, then wax the knot. These precautions will help the knot slip right

A hay elevator is something you should buy or borrow. This close-up of the loft end of the elevator gives you an idea as to how it works. The chain continually rotates around the frame, and its spikes (top of the picture) grab the bales and move them along the rail. The bale is deposited at the top where a helper moves it out of the way of the next one.

Right: Hay trailers have front-axle steering, which means the front wheels will turn as you pull the trailer around a curve or turn. Here you can see the tie rod and king pin, much like an old truck axle. This is for great maneuverability when pulling the trailer, but it makes backing the trailer really tough. Be sure to practice before you need the skill or you'll damage a barn, the trailer, or your pride.

Below right: One problem with the spikes that grab the bale is that they can chew up the bale as it comes off the end. Here you see a close-up of the loft end, but this shot details the bale lift loops. These metal loops force the bale off the spikes before the spikes start going back under the elevator. This prevents the spikes from chewing up the bale. If your elevator doesn't have these types of loops or something similar, have a local fabrication shop add them for you.

This picture show proper elevator loading technique. Stay away from the bottom of the elevator because the chain will grab boots and their laces. Load from the side and away from the motor. Wait a few seconds before loading the next. Most motors sized properly for 14- to 20-foot elevators won't be able to raise more than two or three bales at a time without overheating. Of course, if your elevator is short, this may not be a concern.

through the baler's tensioning and knotting mechanisms and result in no downtime.

Keep a can of WD-40 or other water-displacing lubricant on the tractor. Spray the balers knotting mechanisms frequently and spray any unpainted, mild steel that has to be left out in the field overnight. This will prevent flash rust caused by dew.

Don't stop harvesting because you broke tedder or rake tines. These implements will operate fine with a tine or two missing. Replace them after the harvest.

Operate the cutting head of your mowing equipment as close to the ground as possible. Leaving the grass tall, like you might a lawn, makes raking less efficient, encourages many forage and hay species to grow coarser leaves and stems, and unless you anticipate a dry spell or drought, the extra length does not help plants either.

Most haying equipment, especially mowers and balers,

have shear bolts that are designed to break so that the more expensive parts won't break. Your baler undoubtedly will have at least one. Keep some in stock because you will likely break a few a year.

Consider using the new sickle bolts instead of sickle rivets when making field repairs on broken sickle sections. These are much easier and faster to use in the field than carrying around a rivet press and riveting sickle sections.

Don't forget the classified advertising sections of your state's agriculture department's website and/or magazine as a place to sell your hay.

Keep all equipment is the best shape possible. Equipment in excellent repair is safer.

If you get cell phone reception in your field, carry a cell phone with you so you can contact help if you need to, but never use a cell phone while operating equipment.

CHAPTER 5
TRACTOR AND IMPLEMENT MAINTENANCE

Small-scale haymaking is full of compromises. One of the large compromises we make is that we can not use new equipment. It just doesn't make economic sense. The fact that used equipment needs much more TLC and attention implies that we have to give maintenance, inspection, and repairs our full attention and not ignore these important tasks.

Try your best to locate manuals for your implements. Among the dealer of the equipment, the manufacturers themselves, tractor and farm equipment manual dealers (see appendices), and online auctions, you should be able to find virtually any manual you need. In the absence of a manual, try to identify a newer model of your implement that is very similar. For instance, the New Holland 60 series model baler is not much different than the New Holland 70, so obtaining the New Holland 70 baler manual would suffice in a pinch.

EQUIPMENT INSPECTION AND LUBRICATION

Unfortunately, maintenance and inspection are the types of tasks that are often postponed. In farming they can't be ignored, because life, limb, and valuable property may depend on how well the machines operate. Discovering problems, broken parts, needed adjustments, and safety issues must absolutely be found ahead of time—not afterward when parts go flying. Another task that goes hand in hand with inspection is lubrication. I always inspect and lubricate the equipment at the same time to save time.

Pre-use inspection of your baler starts with the pickup tines. Are any missing, bent, or broken? Spin the baler by hand to verify that none of the tines interfere with the flat band found between them. These bands keep the hay near the tips of the tines and clean the hay from the tines as they clear the bands at the feed mechanism.

Equipment inspection also means inspecting the twine. Here is some twine that has some mold on it. Typically this twine will break, as it did here, with the slightest provocation. Cut this section out of the spool and reset the knotters.

Since tying the wrong ends is a common mistake, these are the ends to use when splicing twine. The left spool is feeding the knotters. Since we always feed from the inside of the spool, splice using the outside end and tie to the inside end of the reserve spool.

A Quick Lesson on Lubrication

There are three terms that I use in this section—lubricate, grease, and oil. Lubricating means applying any kind of friction-reducing compound, whether it is oil, grease, graphite, etc. When I use the terms grease or oil, I specifically mean to actually apply grease or oil. In hay farming, you will use grease and oil, or you may use a lubricating spray as an oil substitute. There are many reasons for using a lubricating spray instead of oil. It is lighter bodied so it is better suited for when you want to coat an entire item to protect it. An example is spraying a knotting assembly before placing the baler in storage. These sprays prevent rust nearly as well as oil, but are cheaper and easier to apply than brushing oil on a part. You can also use spray when you are lubricating two points that occasionally rub against each other with very little load or force. An example is the PTO

Left: When making a splice knot, the knot must be the same thickness of the twine and no thicker. This knot would never pass the tensioning plate without causing twine breakage or the knot to come undone. Right: This knot will clear the tensioning device and knotting mechanism.

One day I was frustrated with frequent twine breakage. Usually we suspect twine tension or knotter knife adjustment when this happens, but neither of these could account for the breakage. After tracing the twine, I found the culprit—a twine guide where an edge had chipped off and was fraying the twine.

shaft support brace that swings in and out on your implement. A quick spray on these items is all they need. These lighter-bodied lubricants also don't attract as much dirt and chaff as oil and grease, which keeps working parts free of the foreign debris that cause wear. For example, oiling a sickle bar with heavy engine oil is asking for dirt and grit to hang around the ledger plate and sickle sections and dull sections and increase the wear.

Of course, nothing is simple, and greases and oils are no exception. Wheel bearing grease should be used for wheel bearings, but it has too high a soap content to be used on much else. General-purpose grease is the most commonly used grease and is what you should use for virtually everything. Molybdenum disulfide grease should be used on universal joints and slip joint components, such as two halves of a shaft that slide inside one another. Lithium grease is great, and most greases sold as general-purpose grease are lithium-based. As for oils, 20-weight should be used for chains, 10-weight should be used for intricate assemblies, and 30-weight should be used as an antiseize on bolts before you fasten them. As for the fluids to use with transmissions, engines, power trains, and hydraulics, follow your equipment manufacturers' recommendation.

THE TRACTOR

There are three major areas of the tractor you should inspect closely: the powertrain, hydraulics, and PTO. Engine inspection begins with the normal sorts of inspections, such as coolant and oil levels, followed by a full search for any oil leaks. Oil-leak detection is important since it is tough to notice low oil pressure via the gauges because you usually are watching and concentrating on the implement. The best defense against an engine ruined by lack of oil is to make

Before every use, load up with twine. Here are all four spools of twine loaded up and knotted together.

Left: Inspection continues by making sure all twine routing guides are intact. Also double check your routing work to make sure the twine isn't twisted. Right: This is a well-formed knot. My baler forms knots that are different than most balers, but this gives you an idea of what you are looking for: good ends that are neither too short or long, evidence of a clean cut, and loops and knots that are tight but don't look overly stressed. A baler is capable of tying a knot that is as strong and nice-looking as one you can tie. If you wouldn't accept the knot from yourself, don't accept it from the baler.

sure you have a very good sense of the number, origin, and rate of any oil leaks so you will know how often to check and top off the oil.

Next, you should check the electrical system. The battery receives heavy shocks as the tractor is operated through rough fields and is exposed to high temperatures. It is very common for a tractor battery, even a heavy-duty agricultural version, to last half as long as a car battery. Keeping it maintained and inspected is important because there is nothing worse than realizing your battery is dead when you are miles from any other vehicle. Test the battery often and keep the electrolytes up to proper levels if the battery isn't

the maintenance-free variety. You should also test other electrical components on the primary (nonignition-related) electrical circuits such as glow plugs if your tractor is a diesel, lights, and other electrically operated accessories, when you check the battery.

If your tractor has a gas engine, inspect the secondary electrical systems and the high-voltage engine spark control. These inspections include pulling the spark plugs and wires to make sure they are in good shape and that the spark plug wires are routed properly (i.e., they don't cross each other, aren't pinched, etc). Make sure the rotor and rotor cap are clean and in good condition.

Sharpening the knotting knives on a baler requires a bit of patience and care, but isn't difficult. The first step is to remove the knives. The knotting mechanism on most balers will rotate up (after the removal of a bolt or two), as shown above. When removing the bolts for the knives, be careful to catch and keep all washers and shims that come off. Take the knives into the shop and sharpen them with files and sharpening stones. Be sure to get as sharp an edge as you can. Strop the edge to remove the small wire burr created on the edge during the sharpening process. Reinstall the knives and be sure to follow any positioning guidelines. The edge of the knives should be ½ inch from the knotting cam on this baler. This critical dimension is created by adding and removing the shims.

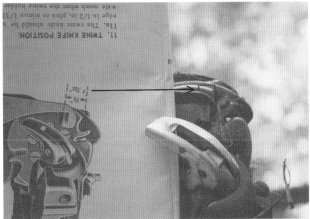

If your tractor has a diesel engine, make sure you regularly service the fuel filters. If your tractor has a water separator, drain it after every use. Make sure your gas tank is full before you head out to the field. Letting a diesel engine run out of fuel is exasperating because getting the air out of the fuel system so it can prime itself again is tough and time consuming on some models. Some tractors have a hand priming lever on the fuel pump for this circumstance, but many don't. The best way to prevent this is to never gamble on how much fuel you have and always keep it topped off. If your air intake filtration is oil-based, make sure you don't overfill the oil. Diesel engines will run on the oil from the intake and redline themselves if the oil bath filter is dramatically overfilled.

There are a few last inspections and lubrications to be done every day. If the tractor has oil cups, oil the generator/alternator. Check the belt tension. A ½-inch deflection is standard across the majority of models. Start the engine and inspect its running condition to look for any coolant leaks. Also look for any looseness, wobble, or abnormal noise from the water pump, belts, and engine-driven components, including the generator/alternator. Lastly, go over the tractor and grease all of the grease fittings and make sure

If you decide to mow with a disc mower, the edge knives should be inspected and sharpened as needed. This photo shows a disc knife and the typical sharpening angle. Notice the mounting bolt. Please be careful reinstalling it to make sure the knife swings smoothly and the bolt is secure.

all the tires have enough air pressure (10-15 psi in the rear tires and 20-25 psi in the front tires is standard).

Next, check the hydraulics. Most haying equipment is self-drafting (the tractor doesn't provide height control). The exceptions include three-point mounted rakes and some mowers. Therefore, if you have those kinds of

Hydraulic hoses that are loose and unanchored are common on used farm equipment. Maintenance includes rerouting and anchoring hydraulic lines as needed to prevent torn hoses and equipment outages. Hydraulic lines should have plenty of slack at the ends to accommodate needed changes in length. Loose ends, like the one shown on the left, should also be placed in a stand-off to prevent entanglement.

equipment, you should closely inspect the three-point hitch of the tractor. However, it is common for hay implements to have a hydraulic cylinder for raising and lowering portions of the implement to facilitate transport of and to adjust the operational height. So the remote hydraulic system of your tractor will need to be routinely inspected. On the vast majority of tractors the same pump that drives the three-point hitch also provides remote hydraulic ability. Checking the level and condition of the hydraulic fluid in a single reservoir is usually all that is needed for both the lift and remote hydraulic systems. Hydraulic systems require absolute cleanliness, so make sure remote ports are kept clean and are covered when not in use. If your tractor is older, look closely at the lines and hoses leading to the remote ports. They are often cracked and leak and might need to be replaced.

Our inspection revealed a broken tedder tine. This particular tine is broken outside of the mount so the tine is still well anchored and isn't a safety hazard. A single broken tine won't change the effectiveness of the tedder, so you can still continue using the tedder if a new tine is not available right away.

Above and next page: These are examples of some of the harder-to-spot grease fittings you must catch during your maintenance. There is one under a cultipacker bearing, right at ground level. Count on at least eight fittings on knotting assemblies, with 10 or 12 being the most common. Some grease fittings found on implements are marked like these tedder and mower fittings, but others are completely obscured unless you are on the ground looking up, such as under the hub of a pinwheel rake or under a tedder basket. Don't forget shafts that slide in and out of the universal joints. They need occasional grease, too.

To check the fluid level of the hydraulic system, there is usually a dip stick found on the main case of the tractor or there will be a small plug that will dribble oil if the fluid is high enough. The dip stick or plug can be found near the lift, which is typically behind the transmission port of the main vase and ahead of the final drive. Often, they share its fluid with the final drive/powertrain of the tractor. In either case, these systems are notorious for building up moisture from condensation so you need to check the consistency of the fluid, too. The hydraulic fluid takes on a chocolate milk appearance because of condensation. When you see this, the fluid needs to be changed. The hydraulics will lift much more smoothly and maintain height control better if you have fresh fluid.

The last hydraulic inspection is the lift test of the three-point hitch. Hook up one of your implements and watch the hydraulics. Does the implement stay still when it reaches its full height or does it move up and down constantly, even if the movement is small? Does the implement fall down very quickly if you turn off the tractor? Does the hydraulic system fail to lift an implement that is heavy, even though the implement is within the specification of the tractor's lifting capacity? If you answer yes to these questions, your hydraulics have seen better days and you may want to consider doing a more in-depth inspection with some additional tests to determine if your system needs to be rebuilt. These issues probably shouldn't keep you from harvesting when you want if they seem minor, but you should address them at your earliest convenience.

Lastly, double-check the PTO system. If the PTO is a direct drive (the PTO only operates if it is engaged and the tractor's clutch is engaged), make sure the engagement lever operates smoothly and easily and the shaft operates without a distinct wobble or noise. Check the seal to see if there are any fluids leaking from the seal at any appreciable rate. If the PTO shaft is live (the PTO shaft operates from engine power that bypasses the powertrain's clutch), check the clutch of the PTO shaft. If PTO engagement is through a two-stage foot clutch near the engine, see if the clutch operates smoothly with a well-defined engagement. The portion of the clutch travel dedicated to PTO engagement and disengagement should be short and without any real opportunity to slip

A part of successful small-scale haymaking involves finding affordable used parts. The cheapest way to obtain parts is to buy an identical implement that is not serviceable to use as parts. This New Holland 273 sits as a parts machine and has a few already missing. If you have ever had to pay for a new replacement needle, you'll realize the wisdom in this approach.

the clutch like you might with the portion of the clutch devoted to the tractor's powertrain. If the PTO clutch is near the PTO shaft itself, it will almost always be of the over-center variety (the clutch engages rapidly through spring pressure as the mechanism travels over the center of the clutch axis). These clutches should have a very distinctive rapid engagement that is best described as a snap. Any softness, hesitation to disengage, or failure to exhibit strong, clean distinct movement warrants removing the clutch for further inspection.

Next, you should inspect your over-running clutch or consider whether you need one or not. An over-running clutch is a safety device that is placed over your tractor's PTO shaft and prevents implements that have a heavy spinning component, like the flywheel on a baler, from continuing

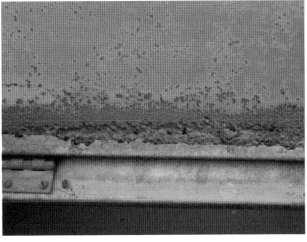

Rust is a cancer of metal, so your inspections should include keeping an eye out for rust. Any metal surface that gets wet but doesn't receive direct sun, or any area that becomes caked with grass dust or fertilizer, which is acidic, will rust badly in no time. This fertilizer box lid is almost lost to rust. Cleaning these metal surfaces and keeping them coated with an oil-based lubricant will help to prevent rust and minimize any rust that is occurring.

Left: A thorough inspection means getting under the hood of all implements. Balers feed the hay into the bale chamber through an auger, feed forks, or a mixture of both. This photograph shows the feed forks' track and mechanism. Make sure the track has a light coat of oil and is free from damage, deformation, or heavy rust. This is a perfect application for a light oil from a spray can, such as silicone spray or WD-40. Right: Here is a close-up of the bottom of a sickle mower's rock guards. They should be inspected from the bottom, checked for cracks, and cleared of dried mud and debris. The guard appears to have a serrated edge, which is actually from a small plate attached to the top of the guard. Inspection of the guard's bottom will show ledger plate wear more quickly and effectively.

to turn the rear wheels of the tractor forward after the clutch is pressed in. Having the flywheel move the tractor forward is an exceedingly dangerous condition that is very common on direct-drive tractors. It will also occur on live PTO tractors if the clutch fails to disengage. If you are unsure whether you need an over-running clutch, err on the side of caution and use an over-running clutch. They are not very expensive and can be used on any tractor regardless of the type of PTO.

IMPLEMENT INSPECTION AND LUBRICATION
Mowers and Mower-Conditioners
Mowers and mower-conditioners will create a lion's share of maintenance, lubrication, inspection, and adjustment. Sickle mowers all share the same design: a pitman rod or offset pulley is driven by the tractor's PTO shaft and moves a long thin sickle bar with sharp triangular sections back and forth over a series of hardened cast or forged triangular sections called guards. These guards may have replaceable sections on them called ledger plates. Wear plates within the sickle bar's operating groove gives a smoother and serviceable surface for the bar. Bar clips (often called knife clips) form the upper restraint for the sickle bar. All of these parts break, wear, become misaligned, and rust. How well the mower-conditioner works is directly related to how sharp, tight, and smooth the sickle assembly is running. Therefore inspect the entire sickle assembly and sharpen or replace each section that needs it, and make sure each clip is properly in place and that each guard and its ledger plate is present and straight. To keep the sickle assembly off the ground, usually three or more of the

Proper equipment care also means proper storage. Harrow rakes should be rolled up. If stored flat and left outside they seem to migrate into the soil and weeds almost immediately, making them hard to find and a hazard for anyone walking by.

guards have heavy bosses underneath. This keeps the sickle assembly out of the dirt, but if mud or dirt collects on or around these bosses, the sickles will not cut properly in the vicinity of the guard. Make sure these guards are clean and functioning.

Inspect the drivetrain of the mower-conditioner, which consists of a PTO shaft, transmission, then a series of shafts or belts that moves a pitman where the sickle bar is attached. The sickle bar will have a cast round end where it attaches to the pitman assembly of the drivetrain. This needs to remain well-lubricated. Make sure the belts are tight and in good condition. A little belt dressing is helpful in maintaining the belts. Make sure the PTO shaft is well-greased and undamaged. The transmission will require occasional fluid changes and the transmission fluid will need topping off from time to time. Most use 80- or 90-weight gear oil, but some will use engine or hydraulic oil. Consult your manual for details on your specific model.

Mower-conditioners are typically self-drafting, but they will have hydraulic height control, so check the hydraulic cylinder for leaks and the hydraulic hose for splits and leaks. As usual, you should be keeping the ends of the hoses covered. These implements are designed so that not all of the weight rests on the bottom of the implement as you travel across the field. Through an ingenious use of heavy springs, the implement has an effective weight much less than its actual weight, which allows the implement to float as you operate it. These springs need to be inspected and possibly adjusted. Your manual should have some specifications for spring adjustment, but generally the springs will be loosened if the mower seems to jump too easily every time it hits a bump. Likewise, the springs will need to be

There are a lot of used parts here. Some parts on a mower are service parts, such as sickle sections, that should be replaced with new parts. Other parts should be replaced with used parts. These pictures illustrate an important approach to small-scale haymaking. New parts can't be justified in most circumstances unless there is no other choice. This approach creates the need for a close relationship with a good implement supply store.

tightened if the mower doesn't seem to float as you travel the field.

The conditioning system needs to be addressed next. Check the bearings for the spools or rollers and make sure that the tensioning springs that keep the rollers pressed tightly together are not rusted to a great extent, are well-attached, and their mountings are secure. Failed mountings or attachment points for the springs are the most common way the rollers' spring loading fails. Check the rubber of your rollers for any sizable nicks or gouges in the rubber. Also, if one of your rollers is steel, make sure any ridges on the steel roller have welds that don't have cracks or other failures. You will ruin both rollers if one of the ridges comes loose and goes through the rollers.

Many pieces of farm equipment are driven by belts. The first picture shows a belt that needs to be replaced. Thread failure on the shoulder of the belt and cracking on the inside are typical symptoms used to judge replacement. Considering the cost of many new belts, regular use of a belt dressing product will keep these belts looking and acting like new. Unfortunately, the owner of this Ford hay rake didn't do this and had to replace the belt. The new belt was extremely expensive and the cost of the new belt would have bought several cases of belt-dressing compound.

Last but not least, check the feeding mechanism. Most mower-conditioners have a chain-driven pickup reel that spins and pulls the uncut hay and forage that is in front of the mower as you travel. This is so the sickle will cleanly cut the hay and will travel into the conditioning mechanism without clumping. This pickup reel tends to become loose on its bearings and the chain is often loose. Be sure to inspect these items closely. Also make sure the reel is intact and doesn't have broken tines. On mower-conditioners without reels, there is usually a bar or roller whose purpose is exactly the same. Make sure the bar is undamaged and lubricated.

Disc Mowers

Disc mowers require less attention, but the attention they require is much the same as mower-conditioners. The important difference is that the disc mower has small knives on the circumference of the large spinning discs. The belts that drive the discs and the disc bearings need special attention. Likewise, these knives must remain sharp. By now you are quickly becoming an expert in PTO shaft inspection, and this implement is no different. Its shaft also needs a quick inspection and greasing. The mower may have a hydraulic lift for traveling down the road and/or for articulating the mower further to the side of your tractor. Therefore, a quick check of the hydraulic system, including the hoses, is in order. While these mowers are less complicated than others, they do rotate quickly, so safety and bearing lubrication are the two main thoughts you will need to keep in mind as inspect and lubricate.

Balers

The baler should be inspected before it is hitched to the tractor and while it is on level ground and the wheels are blocked. Make sure the tongue is supported well. Start the inspection with the tires. They must be evenly pressurized if the pickup head of the baler is to be level, and they also must be pressurized sufficiently (typically 20 psi) to fully and completely handle the weight of the baler on rough ground. Next, inspect under the baler. After double- and

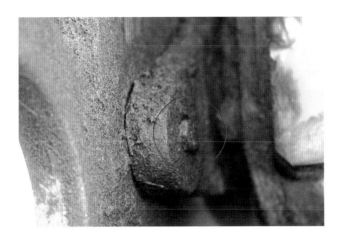

These (above, top right, below) are shear bolts on three common small acreage balers—a John Deere 24T, New Holland 273, and an IH 916, respectively. Shear bolts should be checked every day for looseness and replaced every baling season.

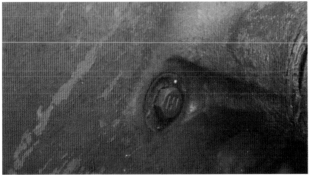

Left: The filler cap on this baler's transmission has a vent tower, which occasionally clogs with rust, wasp nests, etc. Wiggle the cap during your inspection to make sure it is free, and clean it out if it isn't. Right: Don't let rust fool you into thinking you need to replace a part. This sickle section still has clear serrations and no damage, even though it looks pretty rough. Spray it with light oil, and after using it for an hour the sickle section will look brand new.

Left: Sickle hold-downs are often bent and can be broken. This hold-down shows you what a good one looks like. There is very little clearance between the hold-down and the sickle bar and it shows no damage or deformation. Right: Maintenance for haybines and mowers includes spraying light oil on any track, rollers, or bearings that you see that don't have other means for service, such as a grease fitting. One of the feed head rollers in its track is pictured here. Rotate the mower by hand as you spray the track and rollers down with a light oil spray. Do this every time you use the implement.

Keeping a well-stocked supply of maintenance and service parts is a must for any operation, large or small. In addition to the normal assortment of pins, clevis, and shear pins, I like having a full tune-set on hand, a spare fuel pump, a gas tank strainer housing, and various implement parts such as a pitman arm (this one is from a combine), sickle sections, and rivets. Experience with your equipment will dictate what parts wear or break frequently.

triple-checking the blocks that secure the wheels and tongue jack, climb under the baler with a grease gun and spray lubricant. Make sure all of the chains are tight and lubricate them while you are under there. Grease any fitting found under the baler. Inspect the frame for heavy rust, bent or broken braces and struts, and make sure any protective rails (pieces of metal designed to protect the moving parts of the baler) are straight and in place. Often a set or two of twine guides are situated under the baler so check

them while you are under the baler. The guides are usually porcelain and will damage twine if they are chipped, cracked, or broken, so look for any damage.

Check the pickup and feed mechanism of the baler. The pickup head should move freely between its different settings without interference from the tine cleaners or the frame of the baler. All tines should be in place, and replace any missing tines. If replacing isn't feasible and more than one tine is missing, shuffle the tines so that none of the miss-

In small-scale haymaking operations, many things have to be repaired if possible. A cracked carburetor on this John Deere B left the tractor dead in its tracks and a used replacement is very expensive. While most people would assume there is no alternative to replacing it, brazing is one type of repair that is quite easy for any welding shop, and the repair costs about 15 percent of a used replacement carburetor.

This is what happens when you don't examine your baler, including the bale stops, carefully before each use. Bale stops are small blocks designed to stop the plunger if the needles get out of time with the plunger. This prevents the plunger from striking the needles. On my baler a rod that connects to the bale stops lost its cotter pin and disconnected. The next time the flywheel shear pin broke, my needles broke, too.

Occasionally bad things happen to good people. Here a crankshaft pulley broke in half. It took the rest of the day to scrounge up a used replacement and swap it out in the field. I lost half of my hay from this cutting because this delay put me behind in loading it into the barn. In small-scale haymaking, you will lose a few bales of hay to rain unless you live in an arid area. It is just part of the business.

ing tines are side by side or right in front of the other. Next, make sure the feed auger floats up and down evenly and smoothly. If your baler is equipped with feeding forks (the mechanism that feeds hay into the bale chamber), inspect them and check the shear pin to make sure it isn't broken. Make sure the forks aren't bent and any spring is well-attached and not rusted through or broken. As for adjustments, make sure the forks have the reach and clearance you need. The forks are adjusted based on the condition and

A rivet setter and punch is a tool designed to press out sickle section rivets (bottom portion of the tool) and heads new rivets in place after sharpening or replacing the section (top end of the tool). This is a particularly handy tool in the field and shop.

This sequence of photographs illustrates sickle section replacement. The old section was installed with rivets and will be replaced using the newer style section bolts. First, mount the sickle bar on a vise and grind off the head of the rivet. Gently tap the section with a hammer to work it off the rivet. Sometimes the rivet will fall out, but sometimes you will have to punch it out. The next step is to install the bolts. Usually the head of the bolts are up and the nuts are down, but check the clearance on your mower to verify that arrangement will work for you. Sometimes new high-clearance hold-downs have to be installed when switching from rivets to sickle bolts.

Before starting each day, double check your bale length adjustment. Here the locking collar with the set bolt determines the bale length, and every baler has a similar adjustment. Bale tension and condition of the hay influences bale length so the setting will have to vary slightly to keep bale lengths consistent.

volume of hay being fed into the baler. However, it has been my experience that once you find a position that seems to work for your fields and hay in virtually all conditions, it is rarely adjusted after that.

Check the powertrain of the baler, which encompasses the flywheel, crankshaft, plunger, plunger connecting rod, and transmission. The transmission will be oil bathed and is almost always filled with 80- or 90-weight gear oil. Make sure the oil level is adequate. Typically, there are just fill and drain plugs to fill the oil. Check the condition of the connecting rod's bearing. It shouldn't be loose or have excessive clearance. Check the shear pin of the flywheel and make sure it is tight and in good shape. To prevent damage to the baler, it has a clutch mounted on the flywheel that is designed to

let the PTO shaft spin, even if the baler is jammed. This clutch is a spring-loaded set of friction discs and ridged plates. Typically, there is little to inspect, but look for a grease fitting near the flange or hub that needs to be lubricated and check the condition of the springs on the clutch.

The knotting mechanism is difficult to inspect directly. The best indicator of knotter condition is to inspect the knots. About the only thing you can do is a thorough visual inspection when you trip the knotters as part of the threading routine for the twine. Of course make sure it is greased sufficiently and coated liberally with a water-displacing oil, such as WD-40. Make sure all dust, chaff, and other hay debris have been cleaned away. After that, treat the first few bales that you pick up in the field as test bales.

A good, solid, properly sized bale is shown here next to a misshapen and short bale. Not only was the bale length setting incorrect, the lopsidedness came from a dull bale knife and packing fingers that were not properly adjusted.

This is the bale twine tensioning plate. Every baler has one of these and it is very important to the formation of the knot. If it is too tight, the knot may not complete or will break apart easily. If it is too loose, the baler will miss knots or allow the twine to wrap around the knotting fingers and jam the knotting mechanism.

When you do this, get off the tractor and check bale tension manually, make sure the knots are tight, and that the twine is cutting properly.

Lastly, the chute and bale chamber should be clear and rust free. Look for bales and tension plates that are loose and protruding into the bale chamber. These may damage your baler. Cycle the baler by hand and check the bale knife. It should be clean, sharp, and well-mounted. A loose bale knife will create all kinds of damage in your baler. Go back over the entire baler and check all chains for lubrication, that all grease fittings were greased, and that all bolts and shear pins are in place and in good shape. Of course, make sure you have enough twine and that it is properly routed through the guides and needles.

Hay Rakes and Tedders

The inspecting tasks of these two implements are very similar. In fact, you inspect rotary hay rakes (pinwheel and basket rakes) pretty much the same way you inspect tedders. While side-delivery rakes have some of their own idiosyncrasies, their inspection is similar enough that we can cover them all here. First off, inspection and lubrication of rakes and tedders should be done with a certain safety issue in mind. Rakes and tedders, with their array of heavy spinning parts, should never have any loose parts. Loose tines on your tedder become projectiles if they break loose. While these rarely will be thrown high enough to cause an injury, it isn't outside the realm of the possibility either. Plus they can cause damage to both the tractor and implement.

On both of these implements, start with the PTO shaft. Damage to the PTO shafts of tedders and rakes is common. Careful inspection is especially necessary if your rake or tedder is a three-point-hitch-mounted implement, because the shaft tends to rub against the frame on the implement for the hitch during turns that are too sharp. Make sure the two hitch pins on the implement that the lower arms of the tractor attach to are solid and secure. Inspect the universal joints in the PTO shafts and lubricate each one. Make sure the splined end is clean, undamaged, and doesn't exhibit exces-

This is what happens when twine tension is too tight. One of the knots will give way while the other might hold, which will cause the bale to curve around on itself. If this happens too frequently, check to make sure the twine tension is even and proper for both leads of twine. My baler's manual calls for 11 pounds of pull at the tensioning plate. Tie a loop in the twine at the tensioning plate and use a scale and set this accurately. As you get to know your machine, you will be able to tell the tension is correct by feel.

A properly adjusted baler that throws bales that look like this every time is what you want to achieve. Initially, it can be frustrating because it can take some time and practice to get your baler dialed-in, but once you do, your baler should perform nearly flawlessly. At first, you may get up to a 20 percent defect rate, but within a cutting or two, you should be down to 10 percent or less defect rate. After several years with my current baler, I am down to about one to three misshapen bales every 100 bales.

sive wear. There is also a catch pin on the end of the shaft that catches in a groove on the tractor's PTO shaft. This secures the end of the implement's PTO shaft to the tractor. Make sure this pin operates freely and lubricate it with lubricating spray. To accommodate length changes in the shaft as your tractor turns, the PTO shaft is made of two pieces that are slip-jointed. Apply lubricating oil on each

half and work the two halves back and forth to distribute the spray. Make sure the halves slide easily.

The inspection continues with the tines. Are all the tines present and accounted for? Are they bent or misaligned? Fortunately most of these are easy to spot or to repair, but when I inspect my side-delivery rake, I have to watch the rake operate to catch all the tines that are slight-

This is what a shear bolt is supposed to do. After the baler ingested a packing strap, my baler's plunger came to an abrupt stop and sheared the bolt. The broken bolt is on the driveshaft flange.

ly bent. On all rakes each tine is either spring- or rubber-mounted, is extraordinarily long given its diameter, and is flexible. This allows the tine to give as it hits the ground or field obstructions. The tines' rubber will crack, springs will break, or its mounting will bend so be sure to inspect the mounting of the tine as closely as the tine itself. On side-delivery rakes, inspect the clearing bars that the tines run between. The bars can't be bent or else the tines will rub against them.

After inspecting the tines, it's time to lubricate. Grease all the grease fittings you can find and be sure to oil or grease all bearings, bushings, and spots on the implement where pieces bear against or slide past each other. These must be lubricated and kept clean. Wheel position is often part of height adjustment of these implements, so make sure they are mounted as needed for your fields. Lift the implement slightly and grab the wheel by the tire and rock it back and forth to expose roller bearing wear or a worn axle. Most implement wheels don't use tapered bearings and don't have a bearing preload you can check like a car or tractor wheel. Generally speaking these axes and wheel hubs have to be excessively worn before they are unserviceable, so just one check a season is all that is necessary.

Belts should show no signs of cracking, excessive wear, or evidence that the belt is rubbing against something. Turn the implement by hand and look for belt travel and clearance. There are several excellent belt dressings that protect the belt from heat and UV damage. Lastly, make sure any safety shields are in place and any articulating pieces, like the outside baskets of a bat wing tedder, are locked into operating position.

Sharpening

Sharpening is one of those constant tasks on the farm—you just can't get ahead of it. Twine and bale knives on the baler, sickle sections on the mower, knives on the disc mower, various blades on your tools, pocket knives, ancillary equipment (such as your brush hog), and more beckon to be sharpened as all dull quickly. I won't teach you how to actually raise an edge on steel. Instead I will discuss the three sharpening tasks specific to baling hay in a general way and discuss the adjustments of these knives in more detail.

By far the most important sharpening task on a hay farm is to sharpen the knives found in any knotting mechanism. To understand why they are so important and how to adjust the knife afterwards, I will explain how the knotting mechanism tightens the knot and releases the twine. Most knotting mechanisms found on farm equipment made in the last 50 years tie knots using the same general approach. First, the mechanism ties the knot and then it grabs the twine a few inches from the knot and then rotates or flips the device that is holding the knot. This action releases the knot from the knotting device (or at least orients it so the bale plunging action can pull it out) and lays the twine against the knotting knife. The next time the bale plunger compacts the hay, slack is removed from the twine,

the knot is drawn tight, and the twine is pulled quickly against the knife, which cuts the twine and tightens the knot in just one motion.

As you might imagine, all of this requires that the knife must be sharp. The position of the knife is very important. If the knife sticks out too far, the twine will be cut before the knot is drawn tight, which results in knots that fall apart. If the knife is too far back or too dull, the twine won't cut and the knot is pulled out or the stress of uncut twine damages the knotting mechanism. Fortunately the proper setup for the knife will be documented in your manual. In the absence of a manual, set the knife using your best guess after watching a few knotting cycles by tripping the knotter by hand. Try to make sure the knife protrudes far enough into the twine path that the blade will cut the twine easily. Bale a few bales slowly and stop the baler after each bale to inspect the knots and the condition of the twine. Slowly inch the knife back until you get tight knots and clean cuts each time.

The bale knife must be sharpened, too. This knife cuts the hay that is only partially in the bale chamber and is found on the side of the bale plunger. It is a large, mildly hardened piece of steel and because of its size, a grinder is best used to sharpen the blade. To perform a complete sharpening, the bale knife will have to be removed. Because of heavy use and hard objects that occasionally go through the baler, sharpen the edge using a steeper angle than you might ordinarily use when sharpening something. You

should lightly sharpen the knife or dress the edge every time you use the baler. The knife is accessible enough on some makes and models of balers so you can lightly sharpen or dress the edge with the bale knife in place.

Sickle sections are typically no longer sharpened by hand because serrated knife sections cut much better and last much longer than the old straight-edge sections. The serrated edge is impossible to sharpen by hand and keeping new sections on hand is how you deal with a dull or broken section. However, if you prefer straight-blade, hand-sharpened sections (they do work better on very fine grass), I recommend that you keep 6 or 10 extra sections on hand. Keep these very sharp and oiled. Whenever you need a section or two, grab these back-ups, and at the end of the day, take the old ones and sharpen them for the next time you need them.

To sharpen straight-bladed sickle sections, use chisel-sharpening techniques. The back of the blade is kept flat, the leading portion of the edge is sharpened at 30 degrees, and the trailing portion of the edge is sharpened at 25 degrees. This results in good clearance, a sharp edge, and some longevity.

Miscellaneous Equipment

The only other pieces of equipment that are common on small-scale hay farms are hay wagons and hay elevators. Both have inspection, maintenance, and lubrication requirements that are pretty straightforward. Hay wagons

Look for small oiling cups when oiling and greasing. You will often find them on any low-speed application, such as distributors, generators, and water pump shafts on older Farmall tractors.

haul more than 8,000 pounds on a regular basis so make sure your maintenance routine includes inspection of wheel bearings and their preload adjustments. Also make sure that you keep an eye on any wood decking. The inspection/lubrication requirement for hay elevators is likewise straightforward. The idler gear for the chain drive is located at the end of the ramp frame. This idler is under tremendous spring pressure to handle the shock loads of hay bales being thrown onto the ramp. If the chain comes off the idler, it is very hard to get back on. Keep the chain maintained and oiled because worn chains tend to exasperate the problem.

REPAIR AND RESTORATION OF IMPLEMENTS

This is a tough subject to tackle in this book because this is a not a how-to-restore-equipment book. However, the economic realties of small-scale haymaking require that at least some or all of your equipment on your farm need some measure of work. For instance, there is no way my small haymaking operation could justify purchasing even a used side-delivery hay rake that was in field-ready condition, so I had to restore one that was on its deathbed.

If you have the chance to acquire a rusted hulk to restore, keep a few points in mind. You should only acquire implements that are complete. If anything except commodity parts like tires and tines are missing, pass on the implement. You will quickly run past its worth if you are missing significant parts in their entirety. Even small implement tires are expensive. Factor this in when running the

numbers in your mind as you consider your next restoration project. You must also be realistic. If there are no sandblasting companies in your area, are you sure the implement will ever receive that coat of paint it needs? Make sure those tines that look standard and widely available really are. The tines on a tedder-rake I bought turned out to have proprietary tines and the company had been out of business for years. After losing a few tines the tedder-rake was, for all intents and purposes, out of commission.

An implement's restoration starts with necessary repairs, which are typically new tires, sandblasting, and painting. Be aware that tire size is very important and that old tire sizes and new tires sizes don't always agree, even if they are numerically identical. For instance, on the latest hay rake I restored, the tires that came off were very old and were marked as 5.00x15 tires. I put new 5.00x15s on it, but I couldn't lower the rake low enough to rake. After comparing the old and the new I found that the tires were very different in profile and overall circumference, even though the sizes were identical. Save the old tires and compare against new ones.

At this point make the repairs necessary to get the implement field-ready. Replace safety shields, tines, sickle sections, shear bolts, missing bolts, and straighten bent metal, change fluids, and swap hoses. Notice that I haven't mentioned any sandblasting or painting yet because field use is bound to reveal other items that need attention. It will also expose adjustment and operational issues that in turn require additional work. Once you have used the implement

When inspecting equipment, look carefully at the spring mounts. Over time they tend to pull, stretch, and weaken their anchoring points. This eye bolt is showing signs of fatigue at the eye and needs to be replaced.

a few times and exposed areas that need attention, take the implement back to the shop and address those concerns. When you are finished, take the implement to a sandblaster and have the entire implement sandblasted. I also recommend that you have the sandblaster paint the implement if possible, so the bare metal isn't exposed to the elements. After the implement is sandblasted and painted, you are finished and have a great new implement for just a few hundred bucks in sandblasting, painting, and some minor parts.

ADVANCED REPAIR SKILLS

There are certain mechanical and general farm skills you will need to acquire as you continue haying over the years. One of those skills is welding. Haymaking operations will require some type of welding at least once every other year or so. Welding is the only way to fix broken steel. The reason most farmers do this themselves is cost and convenience. Professional mobile welders who come to the farm to repair equipment are expensive, and carrying the implement to a welding shop can be inconvenient, not to mention there may be issues with timing. Simple stick or 110-volt MIG welders cost a few hundred dollars and can perform 80 percent of the repairs you will need. If you spend a few hundred more, you can get a MIG unit that will do virtually all your repairs. Of course, it's your call, but I would definitely wait a year or two before committing to the purchase.

Learning advanced mechanical repairs will also be very handy on the farm. A perfect example is a diesel repair I had to make recently. While I could not rebuild the fuel injection pump, having a mechanic come to the farm to remove the pump, send it off to be rebuilt, and then come back to the farm to install the pump and time it to the engine would have cost hundreds of dollars more than the project actually did cost. Removing the pump myself, sending it to the rebuilding shop, and installing the pump and timing it myself saved the mechanic's charges and the slight markup on the rebuilder's services. This particular task was not difficult, did not require special tools, and didn't require scheduling visits with the mechanic or taking the tractor down to the mechanic's shop. I didn't just save money, I also saved time and trouble. These kinds of skills can be acquired through any community college, a little practice, and a good service manual for the tractor or implement.

TIPS AND TRICKS

Hydraulic Complaints

The hydraulics of the tractor are difficult to diagnose without proper tools and training, but some common complaints lend themselves to owner-operator troubleshooting. Before you do anything at all, I insist that you change your hydraulic oil, and make sure to use compatible hydraulic fluid. The vast majority of hydraulic system complaints I have run across are due to improper fluid, water-logged fluid (particularly common in transmission-case hydraulic systems like the Ford and Ferguson tractors of the N and TO-20/TO-3x series), or low fluid levels.

The hold-down, knife, or clips on a mower are important in creating an efficient, clean, cutting action. Make sure

all these clips are in good shape and hold the sickle sections against the guards before you assume sections are dull.

Cleaning Greasy Parts

Occasionally you will run across the need to thoroughly clean very greasy, nasty parts. By far, the best way to clean them is to dip them is a warm, lye-based caustic solution. This stuff will strip any trace of grease and old paint and leave the part looking like new. To make your own cleaner, simply mix 12 ounces of granular lye to 4 gallons of water. Place the solution in a large steel drum that can be securely closed when not in use and gently warm—do not boil! Here are some very important safety tips for making this solution:

- Always add the lye slowly to the water. Never add water to the lye. Lye and water are highly reactive. By adding water to the lye (especially if you do it slowly), you will create a very strong chemical reaction that may cause the solution to bubble and fume, leaving you covered in a very dangerous liquid that will give you serious chemical burns.
- Always wear caustic-resistant gloves, eye protection, and clothing that covers all your skin any time you are around lye.
- The solution will dissolve some types of metals, especially aluminum and bearing materials. Make sure you don't put anything in the solution that isn't exclusively steel or iron.
- Lye is sold in grocery stores and is inexpensive.
- Large plastic tubs with lids make great inexpensive tanks, but can't be warmed.

Stuck Fasteners

I can promise you that if you spend enough time around haymaking equipment, you will one day run across a bolt, screw pin, or some other part that will not come off. Even after trying your best and using all your might, you will find a bolt rusted into place that won't budge. First of all, don't create extra work by rounding off the bolt or nut or grinding out the screw slots. You can remove the stuck part, and the methods are actually straightforward and obvious once you take the time to think it through.

The various solutions to remove the fasteners using common methods and tools fall into two categories: a different tool and patience, heat, and oil. The first approach realizes that we may have chosen poorly when initially grabbing a tool. It recognizes that it does matter how thick and long the blade of a screwdriver is if we expect to be able to extract it without stripping the screw slot. Sometimes just finding the right combination of tools will do the job. Sometimes the right tool isn't the right tool to use. For instance, a cotter pin rusted into place will stay put if you try to use a cotter pin puller, but a pair of pliers used with a gentle rotating force will work wonders. Another note: a hand-held impact wrench often removes screws and small bolts better than a typical screwdriver or wrench. They also make short work of breaking those parts and fasteners, so be careful.

The patience, heat, and oil approach can be taken at face value. The method consists of applying heat and oil and patiently waiting for it to work. This works because fasteners usually become stuck from of the formation of rust, corrosion, or galling between the fastener and parent material. By alternatively applying heat and oil you can often

Side-delivery rakes that will also ted hay will have a gearbox like this one to select between tedding and raking. This gear box also has a setting for fast or slow raking.

Many rakes have a spring-loaded clutch system to prevent rake damage. This system uses a heavy spring, as seen here, to keep two ridged plates together. Any field obstruction that might jam and break the rake would then overcome the spring tension and allow the tractor to spin the PTO shaft without breaking the rake. This is an older style but is similar in overall design and operation to newer ones.

break the seal this corrosion has between the materials and free up the seized part. The oils used in these methods are collectively called penetrating oils. They are light-bodied and, depending on the circumstance, can be especially effective or utterly useless. To use this method, you apply oil, heat, occasional blows from a hammer, and patience. An acetylene/oxygen welding outfit provides the best heat for this, but a MAPP or plumbers' torch can work well, too.

Paint

As paint fails, it does so by oxidizing and becoming porous, which allows rust to start forming underneath it. Lubricating spray such as WD-40 will also help protect paint that is starting to fail. A quick coating of WD-40 will help arrest the failure of the paint and give you at least a few more years before you have to paint.

INDEX

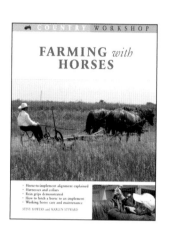